ALTERNATIVE ENERGY RESOURCES:

The Quest for Sustainable Energy

ALTERNATIVE ENERGY RESOURCES:
The Quest for Sustainable Energy

Paul Kruger

Stanford University

WILEY

John Wiley & Sons, Inc.

Copyright © 2006 by John Wiley & Sons, Inc. All rights reserved

Published by John Wiley & Sons, Inc., Hoboken, New Jersey
Published simultaneously in Canada

For general information about our other products and services, please contact our Customer Care Department within the United States at (800) 762-2974, outside the United States at (317) 572-3993 or fax (317) 572-4002.

Wiley also publishes its books in a variety of electronic formats. Some content that appears in print may not be available in electronic books. For more information about Wiley products, visit our web site at www.wiley.com.

Library of Congress Cataloging-in-Publication Data:
Kruger, Paul, 1925–
    Alternative energy resources: the quest for sustainable energy/Paul Kruger.
        p. cm.
    Includes index.
        ISBN-13:978-0471-77208-8 (cloth)
        ISBN-10:0-471-77208-9 (cloth)
1. Renewable energy sources.   2. Energy policy.   I. Title.

    TJ808.K78 2006
    621.042—dc22

                                                                            2005021362

Printed in the United States of America

10 9 8 7 6 5 4 3 2 1

TABLE OF CONTENTS

LIST OF FIGURES

LIST OF TABLES

FOREWORD

Dr. Alan C. Lloyd
Secretary
California Environmental Protection Agency
Sacramento, CA

The last few years have rekindled our interest in energy supplies, centered on various perceptions of availability and sustainability of resources, impending global demand, and the dynamics of U.S. policy in the Middle East due to fear of risks that loom behind new specters of horror, terrorism, and natural disasters. The perceived causes of these adversities are unrelated and hardly close to any semblance of agreement.

Our fears are overshadowed by hope, of the promise of what is achievable with available new technologies such as fuel cells and renewable sources of energy. Such a foundation of hope makes this book by Professor Kruger useful and timely, and important for students to keep on their active bookshelves.

The book is written around a central theme, the never-ending human quest for abundant energy, starting with the three axioms of that quest. The author surveys the status of current and developing energy sources. The text focuses on the three major large-scale energy resources for the exponentially growing demand for electric power and concludes with a section on hydrogen, the current fuel alternative for transportation and electric supply fuel cells. The three chapters devoted to the potential of hydrogen highlight the level of need for this alternative energy source as a fuel, the time period for impact on a scale such that societal benefits are realized, and the available primary energy resources that the author thinks could supply the required energy. The modeling in this area illustrates the consequences of delay in beginning a transition period. The depth of numerical data in the book provides sufficient information so that one can assess what should be an appropriate transition goal.

In this book Professor Kruger updates, in a text format, the alternatives for energy sources. He covers both available and impending resources, representing updated technical and economic information. He also discusses in the various sections environmental benefits and the implications for health. These benefits may be more quantifiable as the costs of chosen directions have become more explicit during the last several decades.

This book is also a useful reference for policy makers who are engaged in developing directions, and in representing information that may already be known. Thus it will be a useful reality check to ground hopes of visionary advocates. Such grounding is required not merely to avoid chasing wrong directions, but also to reduce the costs of not taking or delaying the appropriate steps to reach the end goal.

Dr. Kruger, an emeritus professor at Stanford University, has evaluated, written, and taught this subject for over four decades. His work on energy and the environment influenced and helped our work in the South Coast Air Quality Management District through the 1990s, when we were battling local smog and ozone emissions in the Los Angeles area and looking at the advancement of the cleanest possible fuel technologies.

In that context, Dr. Kruger's work at the time and energy impact of changing into new technologies is pertinent. Most economic evaluations tend to be static, comparing numbers to numbers at a point in time and one case to another. The relevant point in advancing policy and technology is the dynamics of the macro-level, impact of a new technology, the way in which the energy needs will change at a macro-level and what the delay and cost of not starting now can be. It is one thing to talk about hundreds or even thousands of alternative fuel cars such as fuel-cell hybrids, but they are only significant in a region, or a country, like ours, where millions of vehicles are in use. Since we do not want to make policies for small-scale applications, the dynamic modeling done by Dr. Kruger for hydrogen fuels on national and worldwide scales is worth noting. You may arrive at different conclusions from those of Dr. Kruger regarding energy needs, but he leads you to consider the models and to think about other more substantiated conclusions.

We, in my organization, are doing that. I hope you do too.

ALAN C. LLOYD

Sacramento, CA

PREFACE

Undergraduate students, whose education occurs in a narrow age distribution centered on 20 years, generally will experience a professional career with many changes until they retire within a somewhat broader age distribution centered on 70 years. During this 50-year period, one prominent change will be a shift in the continuous human quest for abundant energy, including a major change in transportation from internal-combustion engines that burn petroleum-derived fuels (with an increasing rate of resource depletion and continuing emission of air pollutants) to engines that employ newer technology (such as fuel cells) and new transportation fuels (such as hydrogen, with assured abundance for large-scale production and with negligible emission of air pollution). A worldwide change to hydrogen fuel will allow a logical transition to a dual energy carrier system in which utilization of electricity can be focused on stationary power applications (e.g., residential, industrial, and commercial sites) and hydrogen fuel for transportation applications. With large-scale infrastructure, these two energy carriers can be interchanged readily as needed: electricity converted to hydrogen by electrolysis and hydrogen converted electricity by oxidation in fuel cells.

This book resulted from course notes that were assembled to introduce to freshmen and sophomore undergraduate students and mature adults in continuing education courses the potential for understanding, and possibly getting involved in, the development of hydrogen as a large-scale energy fuel at an early stage in the estimated 50 years required for the technology to be used globally. The text develops a broad picture of this transition in three parts: (1) an analysis of the background of the human quest for abundant energy, (2) the growth in awareness of a world with finite fossil fuel resources and a finite capacity to absorb large-scale waste products that pollute a fragile environment, and (3) the potential improvement that can result from the transition to hydrogen fuel from fossil fuels in transportation.

The text is aimed at undergraduates (and mature adults) in the hope of giving those 20-year-old students an early appreciation of the opportunity to achieve a sustainable world energy supply with acceptable environmental impacts both globally and locally. A wide range of professional talent will be needed to develop the technical and social infrastructure for abundant energy with a minimal regulatory need to curtail freedom of choice.

PAUL KRUGER

Stanford, CA, 2004

1

HUMAN ECOLOGY ON SPACESHIP EARTH

1.0 INTRODUCTION

The search for useful energy by human population of the world may be regarded as one of the constants of recorded history. Another historical constant is the desire for national populations to live together in a clean and safe environment until they are threatened by other national populations when even more energy is needed. A third constant is the irreversible path of development of better ways to generate the supply of energy needed to make societal life safer and more comfortable. This continuous quest for abundant energy may be expressed in three axioms that describe the constancy of discoveries that allow people to do better rather than doing without. The three axioms examined here are as follows:

1. At any given growth rate of the population, total energy consumption will grow at greater rate.
2. Fundamental human goals include both the desire for abundant energy on demand and a clean and safe environment.
3. The future of humanity will continue to follow a one-way and irreversible path.

The conclusion drawn from this examination is that until a more useful higher-specific-energy source than controlled nuclear fission is discovered (such as controlled nuclear fusion), nuclear power eventually will become the preferred energy source for the world's population.

1.01 Axiom 1

Humans have, for comfort, ease, and profit, progressed historically through a series of increasingly efficient (higher specific) energy sources—from humans (self, family, slaves, employees) through animals (oxen, camels, horses) to machines (water, steam, electricity, radiation)—at a continuously increasing rate of consumption of energy per unit of "useful" work. It follows, therefore, that at any given growth rate of human population, total energy consumption will grow at a greater rate.

The rate of change in the population of the world has accelerated continuously through human history. Figure 1-1 shows the history of world population over the 2-million-year period from the Old Stone Age through modern times [1]. This figure emphasizes the rapid acceleration in growth over geologic time that started after the plague epidemic in the Dark Ages.

A more modern history (a picture of long-term world population growth from A.D. 0 projected to 2050) is shown in Figure 1-2, using data compiled by the United Nations [2]. Here the rapid acceleration starts from the time of the Industrial Revolution.

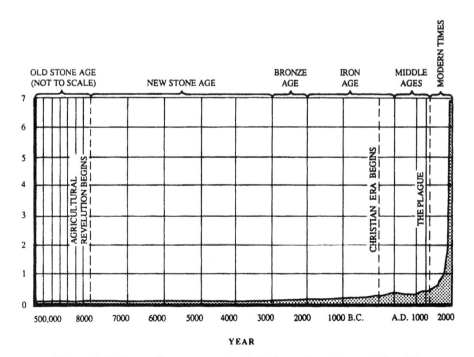

Figure 1-1. Two-million-year geologic history of world population [1].

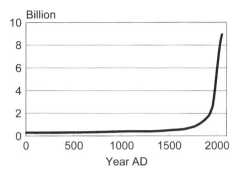

Figure 1-2. Long-term world population growth from 0 A.D., projected to 2050 [2].

The most recent period of history, from 1750 projected to A.D. 2050 by the United Nations [3], is shown in Figure 1-3. The projection shows a decline in annual increments from more than 80 million people per year in the 1980s to fewer than 40 million people per year by 2050. With this steep rate of decline, the world population is expected to reach 9 billion people in 2050 from a population of 6 billion people in 2000.

The acceleration in population growth can be noted from United Nations [2] data in Table 1-1 that show the time interval needed to

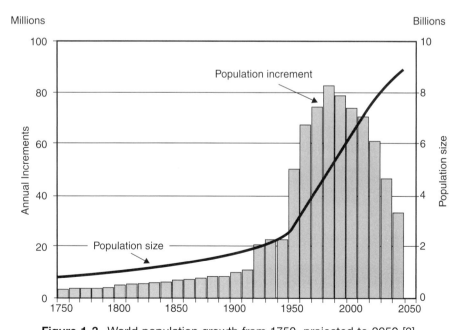

Figure 1-3. World population growth from 1750, projected to 2050 [3].

Table 1-1 Acceleration of world population growth [2]

World Population (Billions)	Year Achieved (A.D.)	Time to Add Last Billion (Years)
1	1804	—
2	1927	123
3	1960	33
4	1974	14
5	1987	13
6	1999	12
7	2013	14
8	2028	15
9	2054	26

increase the population by 1 billion people starting from 1 billion in the year 1804. The lower part of the table shows the deceleration expected by the United Nations on the basis of the declining annual rate projected through 2050.

The corresponding increase in energy consumption as the population has grown can be observed over the geologic and modern time periods. Table 1-2 lists the data for the geologic time period, compiled by Fowler [4] and the United Nations [2]. The population from 5000 B.C. through A.D. 2000 is given in billions with the mean annual growth rate (m.a.g.r.) in percent per annum (%/a) for each successive time period. The energy consumption data are given in kilowatt-hours per capita per day, also with the mean annual growth rate per successive period. The value of 2.9 kWh/cap-day corresponds to the basal metabolism of 2500 kilocalories per capita per day assumed to be needed by prehistoric people.

Table 1-3 lists the world energy consumption per capita for the modern period from 1900 to 2000. The population data are from the United

Table 1-2 Growth in energy consumption from prehistoric times [4]

Period (Date)	Population (Billions)	m.a.g.r. (%/a)	Energy Consumption (kWh/cap-day)	m.a.g.r. (%/a)
300000 B.C.			2.9	—
100000			5.0	<0.001
5000	~0.1	—	9.4	<0.001
0 A.D.	0.3	~0.04		
1850	1.3	0.08	12.0	0.004
1980	4.4	0.94	51.0	1.1
2000	6.0	1.6	230	7.5

Table 1-3 World energy intensity 1900–2000

Year	Energy (Quads)	Population (Billions)	Intensity (MBtu/cap)
1900	22	1.65	13.3
2000	400	6.05	66.1
Increase	18×	3.6×	5×

Nations [2], and the energy consumption data are from the Energy Information Agency of the U.S. Department of Energy [5]. The energy units are given in quads (10^{15} Btu), and the calculated energy intensity is given in million Btu per capita (MBtu/cap). These energy units are reviewed in Chapter 2. The significant observation is the fivefold increase in energy intensity with an 18-fold increase in energy consumption compared with a 3.6-fold increase in population over the twentieth century.

1.02 Axiom 2

Fundamental human goals include the desire for (1) a *pleasant habitat,* defined here as a clean and safe environment, and (2) a life of *comfort and ease,* defined here as abundant energy on demand.

Axiom 2 implies that everyone is an "environmentalist." Most people on earth understand that work (the expenditure of energy) is required to build and maintain a *pleasant habitat* that includes food, shelter, clothing, and the other physical necessities of life. Most people also understand that they do not have to work at all times, that there must be times of rest for *comfort and ease* to enjoy the spiritual aspects of life, such as art, music, and the society of families and other individuals. Most people understand that a pleasant habitat is a clean and safe habitat and expend the energy needed to keep it that way. They generally do this in the context of a life of maximum comfort and ease by reducing personal toil to a minimum (e.g., the energy expended to earn enough for the basic necessities and to maintain physical fitness at a perceived standard). This reduction in personal toil is seen clearly in the rapid adoption of labor-saving inventions such as chain saws, microwave ovens, and electric toothbrushes.

People in large numbers have additional needs for a clean and safe environment. These groups include communities, industry, government, and regulatatory agencies, each of which requires a common expenditure of additional energy. The community congregation of people is illustrated with data from the U.S. census of 1970 on population density in Table 1-4. The table shows that more than 50% of the population lived in fewer than 100 cities. The 1970 census also shows that about 20% of the population lived in 0.1% of the area of the United States.

The trend toward community congregation over a human generation (approximately 25 years) can be estimated by comparing the U.S. census data in Table 1-4 for 1970 with the same data for the year 2000. This is left for the reader to do as a "deeper-look" exercise.

The human quest for abundant and clean energy can be illustrated by the sequence of rapid changes in preferred fuels from before the Industrial Revolution. This quest has been noted by Cannon [6] in a table titled "Moving away from Carbon toward Lighter Fuels" and is expressed in Table 1-5 as the carbon/hydrogen (C/H) ratio. The steep decline from 90% to 0% is shown in Figure 1-4.

1.03   Axiom 3

The history (and future) of humanity follows a one-way and irreversible path.

This axiom implies that civilization as we know it today was developing continuously as the population of the planet increased slowly and remained a small fraction of the sustainable carrying capacity of the earth. The axiom thus is somewhat difficult to defend in that occasional severe events such as wars and disease epidemics have not stopped the growth of civilization by greatly affecting the growth rate of world population. The axiom might be defended by suggesting that there are not enough caves in the world for today's population to revert back to a caveman form of civilization. The possibility does exist, however, for a major fraction of the earth's population to be decimated through a war with "modern technology" (e.g., warfare with weapons of mass destruction) or by a disease that spreads rapidly without means

Table 1-4   U.S. population density in metropolitan areas 1970–2000

Metropolitan Size Range	Number of Cities		Population (10^6)		Population (%)	
	1970	2000	1970	2000	1970	2000
$> 1.0 \times 10^6$	27		68.7		33.7	
5.0×10^5–1.0×10^6	31		22.1		10.8	
2.5×10^5–5.0×10^6	39		13.7		6.7	
Total	97		103.5		51.2	
Total population			204.0	274.6		

Table 1-5   The trend in the C/H ratio in fuels

Fuel	Physical State	Atom % Carbon	Atom % Hydrogen	C/H Ratio
Wood (dry)	Solid	90	10	9.00
Coal (mean)	Solid	62	38	1.63
Oil (mean)	Liquid	36	64	0.56
Octane (C_8H_{18})	Liquid	31	69	0.44
Methane (CH_4)	Gas	20	80	0.25
Hydrogen (H_2)	Gas	0	100	0.00

of controlling it. Even then, if only a few people had to start over they would not have to start from "scratch" in rebuilding the irreversible path of civilization as long as the technical literature remained available.

The technical aspect of the history of humanity is more readily discernible. The development of energy utilization through science and engineering followed a well-established path during the advance of civilization from the discovery of fire and the wheel to the discovery of fission of uranium isotopes and fusion of hydrogen isotopes. An inherent problem in human nature is the almost simultaneous application of each new energy discovery to both civil and military objectives. The history of warfare has been well described (e.g., by Coblentz [7]) and is not considered here. Progress in the development in energy resources also has been described. A convenient way to summarize the history of changes in energy from original solar energy to nuclear energy is shown in Figure 1-5.

The specific energy of fuel, which can be defined as the amount of energy available from a fuel per unit amount of fuel mass (e.g., Btu/

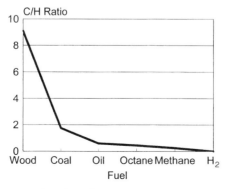

Figure 1-4. The quest for clean fuel as seen in the trend in the C/H ratio.

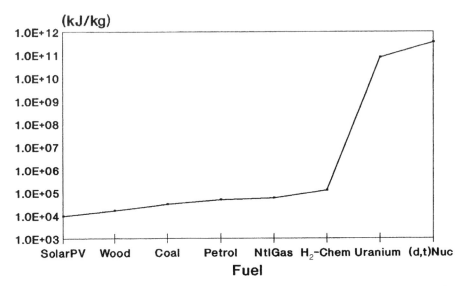

Figure 1-5. The historical sequence of energy utilization as a function of specific energy.

lb in the English system or kilojoules per kilogram in the metric system) is an indicator of the comfort and ease of using a particular fuel. Before fire was discovered, energy on earth was primarily in the form of the heat of the sun and thermal effluents from geothermal and volcanic sources. Unfortunately, the heat of the sun is, by area, too small and diffuse for industrial applications other than providing us daylight for vision, wind power for transportation, and photosynthesis for food production. The solar energy value shown in Figure 1-5 is for solar photovoltaic energy extracted with silicon semiconductor materials that convert solar energy directly to electricity. It has almost the specific energy of wood, which became the fuel of choice when fire was discovered.

The succession of combustible fuels from coal through "chemical" hydrogen shows the increase in specific energy as the choice of fuel changed in accordance with availability. The discovery of nuclear fission energy in the last century provided a choice of fuel about 1 million times greater in specific energy. That power and its introduction as a military weapon have made fission an unpopular choice among a large proportion of the world's population, especially in the case of uranium (and plutonium) fission reactors. The general popularity of thermonuclear fusion as an energy resource is undetermined since its availability (other than as the "hydrogen bomb") has not been achieved in civilian power reactors. However, the ultimate energy resource that could be available, based on today's technical knowledge, is "solar energy on earth," the thermonuclear energy of the sun, which provides the world with its original and continuous energy, using isotopes of hydrogen in properly engineered power stations.

1.04 Philosophical Questions for the Quest

If we accept the three axioms in the human quest for abundant energy, we can wonder whether we are losing our way in this quest. Will energy consumption continue to grow at a greater rate than population, especially if, as forecast by demographers, the rate of population growth will decline over the next 50 years? Will people abandon the goals of a clean and safe environment if the price of dwindling energy resources continues to increase? Will human history continue along its irreversible path?

These questions lead to a central technology question: As human population in cities, nations, and the world continues to grow, even at a declining rate, should we try to reverse the quest for greater specific-fuel-energy technology? These questions also lead to a central social question: As population continues to grow, do we regress (do without) or do we advance (do better)? It is hoped that as we go through the science and engineering of the quest for abundant energy in these pages, some light will be shed on these questions.

1.1   DEVELOPMENT OF HUMAN ECOLOGY

Energy is a fundamental requirement for the sustenance of life. It plays a key role in the development of the earth's flora and fauna, which are termed collectively the biosphere. Obviously, it plays a key role in human ecology, being one of the major factors that govern the well-being of humanity.

The major factors in the long-term growth of civilization can be grouped into the 5 P's, as follows:

1. *Population.* Population drives the need for sustenance; it requires the development of technical means of providing a food supply in the form of agriculture and livestock.

2. *Population density.* Aggregation of populations (communities) drives the need for engineered environments (buildings, roads, bridges, traffic lights) in the form of cities, states, and nations.

3. *Production.* Large populations, with division of labor, drive the need for manufactured product, measured in affluence, in the form of gross domestic product, the total value of goods produced, and productivity (the amount produced per capita per unit time).

4. *Power.* Power is the time rate of doing work (measured e.g., by horsepower [HP] in the English system or kilowatts [kW] in the metric system) that satisfies the need for abundant energy to achieve a desired objective (e.g., driving an automobile at 100 miles per hour or lighting The Strip in Las Vegas).

5. *Pollution.* Pollution, the aftermath of human endeavors, drives the need for a clean, safe habitat; this reflects the importance attached to the health of the biosphere.

1.11   Major Ages in Human History

Figure 1-1 showed the geologic history of world population over the last 2 million years, starting with the Stone Age. Major developments in human history can be associated with the geologic eras, covering long time periods in the early eras and relatively short time periods in the later eras. A summary of significant periods in human history appears in Table 1-6.

1.12   The Biosphere: "Spaceship Earth"

A simple conception of the biosphere is that it is a single, interconnected diversified system for "life on earth." Life on earth depends on the transformation of some of the sun's radiant energy ($h\nu$) that reaches

Table 1-6   Major ages in human history

Age	Period	Significance
Stone	Prehistoric	Food gathering and hunting
Fire	Prehistoric	Separation of humans from animals
Neolithic	B.C.	Long-term growth of civilization
Urban	Dark Ages	Transport and storage of food: cities
Industrial	Eighteenth century	Transfer to mechanical energy
Technological	Twentieth century	Electronics and nuclear power
Future		Solar energy on earth?

the planet into the chemical energy of hydrocarbons through photosynthesis of chlorophyll by the general reaction

$$CO_2 + H_2O + h\nu \rightarrow C_xH_yO_z + O_2 \qquad (1.1)$$

leading to the two major food chain paths. The concept is summarized by the parallel flowpaths

$$[Sun] \rightarrow \begin{bmatrix} photosynthesis \\ of \\ chlorophyll \end{bmatrix} \begin{matrix} [marine\ algae] \\ < \\ [vegetation] \end{matrix} > [producers] \rightarrow [consumers]$$

The concepts of Spaceship Earth and the food chain are described in an early review of the sustainability of resources by the National Academy of Sciences [8]. Chapter 1 by Bates discusses the human ecosystem (Spaceship Earth).

The two major food chains are the aquatic and terrestrial pathways, with producer through consumer components generalized as follows:

Aquatic (marine algae)
   phytoplankton $\Rightarrow$ zooplankton $\Rightarrow$ small fish $\Rightarrow$ large fish

Terrestrial (vegetation)
   plants $\Rightarrow$ herbivores $\Rightarrow$ carnivores $\Rightarrow$ parasites $\Rightarrow$ . . .

These components come into a quasi-equilibrium by the processes of energy flows and material cycling over long periods of time.

Ricker [8] describes the aquatic food chain ("food from the sea") as a complex pyramid of trophic levels of producers and consumers, as shown in Figure 1-6. Hendricks [8] describes the terrestrial food chain ("food from the land"). A parallel question of *who eats whom* shows for plants such as herbs, grasses, and shrubs several first-order consumers such as insects, crustaceans, and herbivores, with recycling decomposers of bacteria and fungi.

1.13 Limits to Growth

The concept of limits to growth sometimes is stated as the rules that "what goes up must come down" and "there is no free lunch." Large-scale worldwide concern about the growth of population, the sustainability of natural resources, and the human impact on the environment

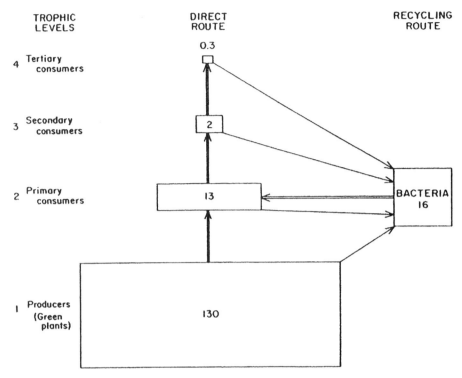

Figure 1-6. The aquatic food pyramid: consumption, production, recycling. From Ricker [8]. Original caption: Simplified aquatic food pyramid, illustrating direct and "recycling" routes for conversion of plant material into animal tissue. Areas of the rectangles are proportioned to the estimated production (*not* the standing crop) of material at each trophic level; production figures are in billion of metric tons of organic matter per year.

may be considered to have started in the social upheavals in the 1960s, but the literature has many references from ancient times. Genesis 1: 28 says, "Be fruitful and multiply, and replenish the Earth, and subdue it." Many lively classroom discussions result from attempts to interpret that quote. In 1798, Thomas Malthus wrote about population as it affects the future improvement of humankind. The revised (and enlarged) revision in 1803 [9] included three key observations:

Population is necessarily limited by the means of subsistence.

Population invariably increases when the means of subsistence increase unless this is prevented by very powerful checks.

These checks, and the checks that repress the superior power of population and keep its effects on a level with means of subsistence are all resolvable into moral restraint, vice, and misery.

Many lively discussions, especially in engineering classrooms, result from attempts to interpret the terms "very powerful checks" and "moral restraint, vice, and misery." Are moral restraint and vice limited to human beings? Does misery (e.g., disease, starvation) affect all creatures?

A large volume of literature on limits to growth developed by the 1970s, as exemplified by the Club of Rome report [10] on the predicament of humankind published in 1972. The major conclusion of the report was, "If the present growth trends in world population, industrialization, pollution, food production, and resource depletion continue unchanged, the limits to growth on this planet will be reached sometime within the next one hundred years. The most probable result will be a rather sudden and uncontrollable decline in both population and industrial capacity (pp. 23–24)." The report provides two solutions to the consequences of its major conclusion: (1) "It is possible to alter these growth trends and to establish a condition of ecological and economic stability that is sustainable far into the future. The state of global equilibrium could be designed so that the basic material needs of each person on earth are satisfied and each person has an equal opportunity to realize his individual human potential." and (2) "If the world's people decide to strive for this second outcome rather than for the first, the sooner they begin working to attain it, the greater will be their chances of success." An update of this book was published in 2004 [11], 32 years later, that discusses some of the processes implied in the first book that are in progress.

As a final example of concern about world survival, the book by Yoda [12] on the world's "trilemma" points out the interrelations between economic development, resources of food and energy, and environmental impacts. The structure of the trilemma is illustrated in Figure 1-7.

1.2 SUMMARY

This chapter made the case that a time in human history has arrived that shows that continued growth in human population, with concomitant greater growth in energy demand, may not be sustainable far into the future. World population is expected to grow from 6 billion in 2000 to about 9 billion by 2050 if the forecast of an annually decreasing growth rate is met. If it is not, world population could grow at the present annual rate and reach 10 to 12 billion people. It was realized in the second half of the twentieth century that limits to growth exist

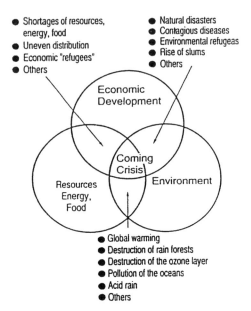

- Shortages of resources, energy, food
- Uneven distribution
- Economic "refugees"
- Others

- Natural disasters
- Contagious diseases
- Environmental refugeas
- Rise of slums
- Others

Economic Development

Coming Crisis

Resources Energy, Food

Environment

- Global warming
- Destruction of rain forests
- Destruction of the ozone layer
- Pollution of the oceans
- Acid rain
- Others

Figure 1-7. Structure of the trilemma: the three major problems threatening world survival. From Yoda [12].

and that if growth continues at current rates, severe problems in the sustainability of needed resources (e.g., energy) could become acute. The following chapters attempt to examine the growth of energy consumption, the sustainability of energy supply, the long-term energy resources available, and the resulting environmental impacts caused on global and local scales, all of which affect the human quest for abundant energy.

REFERENCES

[1] H. Braun, *The Phoenix Project, Shifting from Oil to Hydrogen.* Phoenix, AZ: SPI Publications, 2000.

[2] United Nations, *World Population Prospects: The 1998 Revision.* New York: United Nations, 1998.

[3] United Nations, *The World at Six Billion* (ESA/P/WP.154). New York: United Nations, 1999.

[4] J. M. Fowler, *Energy and the Environment.* New York: McGraw-Hill, 1984.

[5] U.S. Department of Energy, Energy Information Agency, *Annual Energy Outlook 2004,* Report No. DOE/EIA-0383(04) (and earlier years). Washington, DC: U.S. Department of Energy, 2004.

[6] J. S. Cannon, *Harnessing Hydrogen: The Key to Sustainable Transportation.* New York: INFORM, Inc., 1995.

[7] S. T. Coblentz, *From Arrow to Atom Bomb.* New York: Beechhurst Press, 1953.

[8] National Academy of Sciences, *Resources and Man.* San Francisco: W. H. Freeman, 1969.

[9] T. R. Malthus, *An Essay on the Principle of Population as It Affects the Future Improvement of Mankind,* expanded 2nd edition, 1803.

[10] D. H. Meadows et al., *The Limits to Growth,* New York: Universe Books, 1972.

[11] D. Meadows, J. Randers, and D. Meadows, *Limits to Growth—The 30-Year Update.* River Junction, VT: Chelsea Green Publishing Co., 2004.

[12] S. Yoda, *Trilemma: Three Major Problems Threatening World Survival.* Tokyo: Central Research Institute of Electric Power Industry, 1995.

2

THE UNENDING QUEST FOR ABUNDANT ENERGY

2.0 HISTORICAL PERSPECTIVE

The axioms for abundant energy presented in Chapter 1 may be disputed, but the study of natural science has shown that there has been a continuous quest for readily available energy. It is useful to review the history of planet earth in the physicist's frameworks of time and energy. Data compiled from the literature for these two parameters are summarized in Table 2-1.

A comparison of Table 2-1 with Figure 1-1 shows the delay of billions of years between the formation of the earth and the beginning of the 2 million years of human existence, the $\sim$ 10,000 years from the Old Stone Age agricultural revolution to the $\sim$300 years of the Industrial Revolution, and the many "modern" revolutions since the 1905 publication of the equation $E = mc^2$. All this, starting with the human discovery of fire, has constituted the history of the quest. The quest today, with the discovery of electricity and its ever-accelerating development into quantum mechanics, nuclear energy, and electronics, brings us to the next stage of abundant high specific energy.

The values for the energy frame come from the generally accepted model of the energy flow system on earth, as shown in Figure 2-1. The transformation of solar energy reaching the Earth has sustained the planet for at least the last 2 million years and is likely to do so "forever." The distribution of transformed energy components is summarized in Table 2-2. The units of energy and power are described in Section 2.11.

Table 2-1   Planet earth time and energy frames

Time Frame	Years
Age of the earth	$\sim 4.5 \times 10^9$
Human existence on earth	$\sim 2 \times 10^6$
Recorded history	$\sim 5 \times 10^3$
Industrial Revolution	$\sim 3 \times 10^2$
Electricity Age	~ 150
Nuclear Age	100
U.S. Environmental Age	35
Electronic Age	Just started?
Energy Frame	**PW (10^{12} kW)**
Solar energy from the sun	Heaven knows
Solar energy on earth	178
Terrestrial (thermal) energy	0.0033
Lunar (tidal) energy	0.0030

The growth in energy consumption from prehistoric times (see Table 1-2 in Section 1.01), shows the growth from 2.9 kWh/cap-day (the minimum required energy per human) to the current value of 230 kWh/cap-day. This picture of the human quest for abundant energy also has been described by Cook [1] as the changes in the application of energy as a function of the stage of human development. A summary of the data is given in Table 2-3. Key observations from these data are the modest growth in energy expended for the food supply (5× in 10^6 years) compared with the rapid growth (5× in 10^2 years) in energy used for transportation.

2.1   CHARACTERISTICS OF AN INDUSTRIAL NATION

The spectrum of the degree of national industrialization among the world's population includes developed, developing, and undeveloped nations. One of the key attributes that determine in which of these three categories a nation belongs is how the population uses its energy supply. This is one of a number of characteristics that define an industrial nation. Those characteristics can be grouped into three measurable indicators:

1. Energy resources
2. Affluence
3. Trends into the future.

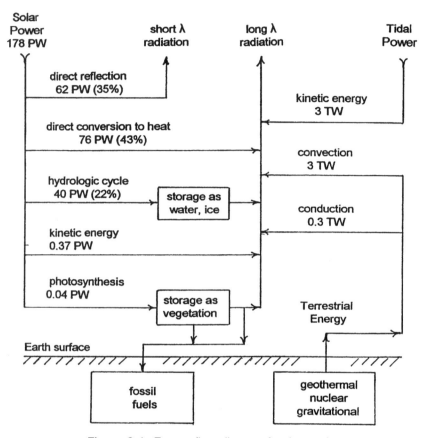

Figure 2-1. Energy flow diagram for the earth.

Table 2-2   Transformation of solar energy

Component	Percent
Albedo (prompt reflection away)	~ 35
Heat (absorption of radiant energy)	43
Hydrologic cycle (water distribution)	22
Kinetic energy (wind energy)	0.2
Photosynthesis (chemical energy)	0.02

Table 2-3 Growth in energy consumption with stage of human development [1]

Stage of Development	Food	Central Heating	Industry and Agriculture	Transportation	Total
Primitive humans (~ 10^6 years ago)	2				2
Hunting (~ 10^5 years ago)	3	2			4
Primitive agriculture (~ 5000 B.C.)	4	4	4		12
Advanced agriculture (~ 1400 A.D.)	6	12	7	1	26
Industrial (~ 1875)	7	32	24	14	77
Technological (~ 1970)	10	66	91	63	230

Energy resources as found in nature come in many forms of solar, lunar, and terrestrial origin. Industrial technology makes it possible to process natural energy resources into a larger variety of forms suitable for various applications. It was noted in Figure 1.5 that the specific energy of chemical and nuclear fuels varies over a large range from 10,000 kJ/kg to 100 billion kJ/kg. One of the indicators for the degree of industrialization is the kinds of energy used in the national economy; the more industrial nations generally use higher-specific-energy fuels. A second indicator is the amount of energy used, which can be measured in units of energy per capita (e.g., kJ/cap) or energy per dollar of gross domestic product (e.g., kJ/$GDP). Here again, the more industrial nations generally consume more kJ/cap and kJ/$GDP. A third indicator is the efficiency of utilization of the energy supply, which can be measured in units of percent useful work achieved per unit of primary energy resource consumed. For this indicator, it is not clear that the more industrial nations consume fuels more efficiently.

Affluence may be expressed as a measure of the standard of living. The indicators may be expressed as degree of economic growth (e.g., $GDP/cap), the degree to which a nation's affluence has an impact on other nations, and the impact on the global environment. These indicators play a lesser role in the quest for abundant energy.

Trends into the future play an important role in the quest. These trends include the social aspects of disproportionate shares of energy utilization (e.g., kJ consumed per capita), the types of fuels used ("capital" versus "income" resources, as described in Section 2.11), and the extent of the choice between convenience and conservation by a nation's population.

2.11 Flow of Abundant Energy

Abundant energy in an industrial nation comes in large "chunks". A review of the terms *energy* and *power* is germane at this point. Energy is defined as the capacity for doing work; power is defined as the time rate of doing work. Energy generally is expressed in mechanical work units. Three major systems are used throughout the world for energy units. The conversion factors are listed in Table 2-4. The three systems are

1. International System (SI), in which the unit of energy is the joule (J):

$$1 \text{ J} = 1 \text{ kg m}^2/\text{s}^2$$

Table 2-4   Energy and power unit conversion factors

Energy	kWh	Btu	Calorie	ft-lb	J (10^7 erg)
kWh	1	3412	860	2.65×10^6	3.6×10^6
Btu	2.93×10^{-4}	1	0.252	778	1054
Calorie	1.16×10^{-6}	3.97×10^{-3}	1	3.09	4.18
ft-lb	3.77×10^{-7}	1.28×10^{-3}	3.24×10^{-4}	1	1.36
J	2.78×10^{-7}	9.48×10^{-4}	2.39×10^{-4}	0.738	1

Power	kW	HP	ft-lb/min
kW	1	1.34	4.44×10^4
HP	0.746	1	3.30×10^4
ft-lb/min	2.25×10^{-4}	3.03×10^{-4}	1

2. Metric (cgs) system, in which the unit of energy is the erg; 1 erg $= 1$ g cm^2/s^2:

$$1 \text{ J} = 10^7 \text{ erg} = 0.239 \text{ cal} = 2.78 \times 10^{-7} \text{ kWh}$$

3. English (ft-lb-sec) system, in which the unit of energy is ft-lb or Btu:

$$1 \text{ J} = 0.738 \text{ ft-lb} = 9.478 \text{ Btu}$$

The respective power units are as follows:

For SI and metric systems: the watt (W):

$$1 \text{ W} = 1 \text{ J/s} = 10^7 \text{ erg/sec} = 3.412 \text{ Btu/hr}$$

For the English system: the horsepower (HP):

$$1 \text{ HP} = 33,000 \text{ ft-lb/min} = 0.746 \text{ kW}$$

Large chunks of energy for national purposes are expressed in two sets of units:

In the United States: the quad (Q)   1 Q $= 10^{15}$ Btu

In the rest of the world: the petajoule (PJ)   1 PJ $= 10^{15}$ J

The range of values in SI units can be illustrated by the comparison between typical daily "environmental" and "technology" energies. The events listed in Table 2-5 cover a range of 10^{50}, a number that can be expressed as 1 with 50 zeros following it. For large chunks of national energy, the petajoule (10^{15} J) and the quad (10^{15} Btu) are appropriate units. The SI prefixes used in this book are *kilo* (k for 10^3), *mega* (M for 10^6), *beva* (B for 10^9), *tera* (T for 10^{12}) and *peta* (P for 10^{15}).

A pictorial way to describe the energy flow in an industrial society was made popular by E. Cook in 1971: the "spaghetti-diagram." Figure 2-2 is adapted from the diagram in Cook [1] showing the energy flow for combustible fuels to generate electric energy for useful heat, light, and work, as well as the energy losses in the process and heat dissipation to the environment and subsequent radiation to space as part of the solar energy cycle.

2.12   Capital and Income Energy Resources

An additional characteristic of industrial nations is the wide choice of primary energy resources and secondary energy forms. Primary energy resources are categorized as capital resources, implying expenditure from a one-time estate without replenishment (nonrenewable), and income resources, implying expenditure of replenished resources without loss of capital (renewable energy resources).

Capital energy resources consist of fossil fuels (e.g., coal, oil, natural gas), which require millions of years for geologic replacement, and

Table 2-5   Typical energy values for environmental and technology events

Daily Environmental Energy	Joules
Outflow from the sun	10^{31}
Inflow to the earth	10^{22}
World photosynthesis	10^{19}
Human energy demand	10^{17}
Niagra Falls conversion	10^{14}
Average per capita use in United States	10^9
Per capita food energy (2000 kcal)	10^7

Technology Event Energy	Joules
Fission of one U-235 nucleus	10^{-11}
Combustion of one C atom	10^{-19}

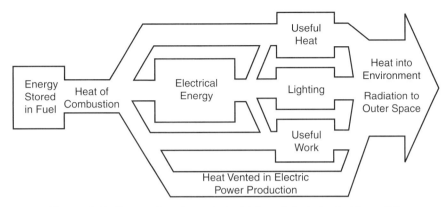

Figure 2-2. Energy flow in an industrial society (adapted from [1]).

nuclear fuels (e.g., thorium and uranium), which, although transformable, are irreplaceable. Geothermal energy, which generally is included among the renewable resources, requires hundreds to thousands of years for thermal energy replacement by conduction and convection of the heat extracted from the geologic deposits of commercial thermal resources. Income energy resources include the potential for lunar energy (in the form of ocean tides) and primarily solar energy (in the forms of thermal, hydro, wind, and biomass energies). Most of the world's renewable energy utilization is achieved in the widespread use of dams on river systems to generate hydroelectric power. Each of these energy forms is reviewed in later chapters.

The quest for comfort and ease of energy utilization has resulted in a growing demand for secondary energy vectors, which are energy forms that are transferred from primary energy resources for convenience, cleanliness, or specific applications. The most notable secondary energy vector is electricity, which is barely more than 100 years old as an industrial form of energy consumption. The book concludes with a review of the potential for large-scale production of hydrogen fuel as the complementary secondary energy vector to replace fossil fuels both for the generation of electricity in stationary power plants and as a carbon-free fuel for transportation in the early part of the twenty-first century.

As an early illustration of the quest for cleaner fuels and convenient secondary energy vectors before we examine the transitions in detail, Table 2-6 shows the fuel mix history in the United States from 1925 through 2000, compiled from U.S. Department of Energy data.

Table 2-6   U.S. fuel mix history (in percent) 1925–2000

Fuel	1925	1950	1975	2000
Wood	7	3	0	0
Coal	65	37	21	32
Oil	19	39	38	21
Natural gas	5	18	32	27
Total fossil	97	97	91	81
Nuclear	0	0	6	11
Renewable	3	3	3	8
Percent of total for electricity	< 1	~ 2	~ 15	38

2.2   EXPONENTIAL GROWTH DYNAMICS

To understand the limitations imposed on the quest for abundant energy, it is helpful to understand the arithmetic of exponential growth. Several types of growth are of interest in regard to both personal and energy considerations. The two major ones are linear growth and exponential growth. Also of interest is the form of saturation growth and, from the concept of what goes up must come down, the growth (and decline) rates of extraction of a finite resource. Each of these four types of growth history is reviewed here before we examine resource statistics.

2.21   Linear Growth

Linear growth may be described as increases by a constant amount per constant time period. If you put $10 under your mattress in the first week in January and add $10 each week for the rest of the year, the amount under the mattress at the end of the year will be $520 ($10 for the first week plus 51 more weeks times $10 per week). The growth of the "under-the-mattress" depository (without earning interest) is illustrated in Figure 2-3.

For an initial amount at time $t = 0$, N_0, and a constant addition of k per constant time period t, the total amount after time t, N(t), can be written as an arithmetic equation:

$$N(t) = N_0 + kt \tag{2.1}$$

In this savings account, $N_0 = \$10$ for the first week, $k = \$10/week$, and $t = 51$ more weeks. In Figure 2-3, k is the slope of the straight line (in units of $/week).

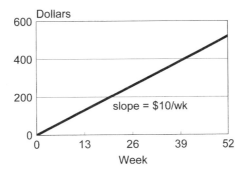

Figure 2-3. Linear growth of an "under-the mattress" savings account.

In more complex linear systems, Eq. 2.1 can be written in differential form (many small increments, dN, in many very short time periods, dt) as

$$dN = k \, dt \qquad (2.2)$$

The solution to this differential equation is

$$N(t) = N_0 + kt \qquad (2.3)$$

which is similar to Eq. 2.1. Here $N_0 = \$10$, and if $dt = 1$ sec, k will be 0.0000165 \$/sec, from \$10/wk (7days/week $\times$ 24hr/day $\times$ 3600 sec/hour).

2.22 Exponential Growth

Exponential growth may be described as increases by a constant fraction of the amount per constant time period. This means that the increase per time period is proportional to the total amount after each time period, and therefore the increase becomes larger with each time step. The differential equation for continuous exponential growth is given by

$$dN = kNdt \qquad or \qquad dN/N = kdt \qquad (2.4)$$

The solution to this equation (for $N = N_0$ at $t = 0$) is

$$N(t) = N_0 \, e^{kt} \qquad (2.5)$$

If, after putting $10 each week for a year under the mattress in linear growth to $520, you put your money in a bank paying interest of 6% per year (k = 0.06/yr) instead of keeping it as $520 under the mattress with no interest, the amount at the end of 10 years will be

$$N(10) = 520 \exp(0.06 \times 10) = \$947.50 \qquad (2.6)$$

Figure 2-4 shows the growth of the $520 compared with keeping the initial year's collection under the mattress for the next 10 years.

2.23 Doubling Time

A useful index of growth is the doubling time (DT), the time period required for a value to grow from N to 2N. This index can be calculated for both linear and exponential growth:

1. For linear growth, Eq. 2.1 becomes $2N = N + k(DT)$, leading to $DT = N/k$.
2. For exponential growth, Eq. 2.5 becomes $2N = N\ e^{k(DT)}$, leading to $DT = \ln(2)/k$.

Note that Eq. 2.5 can be converted to a linear form by taking the natural logarithm of both sides:

$$\ln[N(t)] = \ln[N_0] + kt \qquad (2.7)$$

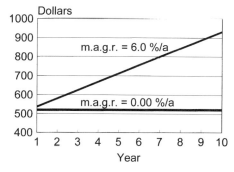

Figure 2-4. Exponential growth at 6%/a compared with no growth at 0%/a.

2.24   Exponential Growth Scenarios

Interpretation of the growth of the parameters that affect the human quest for abundant energy requires the use of three types of exponential growth scenarios: (1) continuous exponential growth, (2) saturated growth to a maximum value, and (3) logistic curve growth.

Continuous Exponential Growth   The equations for continuous exponential growth given in Eqs. 2.5 and 2.7 are illustrated in Figure 2-5 as a function of the number of doubling times in both the exponential and linear form for $N_0 = 1$ and $k = \ln(2)/DT$.

Note that N after the fourteenth doubling time would fall at 16 thousand, well above the chart scale. Thus, the linear form of continuous exponential growth is useful for large-scale growth parameters. However, almost all physical processes that exhibit early continuous exponential growth cannot sustain such growth for a *long* time since the next doubling period results in an addition that is equal to the sum of all the additions that have preceded it.

Saturation Growth   Saturation growth occurs when the rate of growth decreases as the total amount increases, up to a maximum amount, N^∞, which is reached at infinite time. This type of growth is illustrated in Figure 2-6.

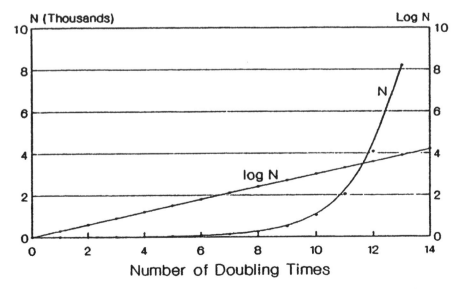

Figure 2-5. Exponential growth as a function of the number of doubling times.

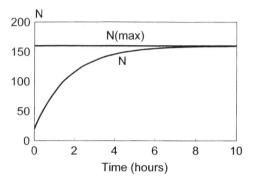

Figure 2-6. Saturation growth to N(max) for $N_0 = 20$, $k = 0.5/hr$, $dt = 0.5$ hr, and $N^\infty = 160$.

The differential equation is given as

$$dN/dt = k(N^\infty - N) \tag{2.8}$$

The solution for $N = N_0$ at $t = 0$ is

$$N(t) = N^\infty - (N^\infty - N_0)\,e^{-kt} \tag{2.9}$$

Logistic Curve Growth   The logistic curve is used to model the integrated production of a finite resource. It is characterized by an early period of rapid (exponential) growth of the production rate, but for a finite resource, the production rate has to reach a maximum value and decline as the integrated production reaches the total (finite) amount of the resource. Logistic curve growth is illustrated in Figure 2-7.

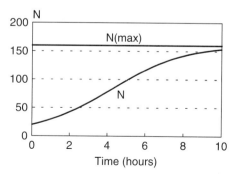

Figure 2-7. Logistic curve for finite N with $N_0 = 20$, $b = 0.5/hr$, $dt = 0.5$ hr, and $N^\infty = 160$.

The differential equation is given as

$$dN/dt = k\,N\,(1 - N/N^\infty) \tag{2.10}$$

The solution for $N = 0$ at $t = 0$ is of the form

$$N(t) = \frac{N^\infty}{(1 + a\,e^{-bt})} \tag{2.11}$$

where a and b are empirical constants and N^∞ is the estimated total production.

2.25 Calculation of Growth Rates by Regression Analysis

Forecasting of future values of a variable from recent history generally is based on extrapolation of the most recent sequence of prior values of the variable that has a well-established historical growth rate. A useful tool for estimating a mean growth rate is the statistical method of regression analysis. This method is especially useful for evaluating the relationship between two variables, for example, chronological behavior of a parameter such as population as a function of time. The data may follow a linear, exponential, logarithmic, or power series relationship. The relationships of interest in forecasting future energy demand are linear and exponential relationships.

Regression analysis provides insight into the coefficients of the relationship and a measure of how well the relationship fits the data. For an annual time series of a variable, the regression (which performed readily in spreadsheet format) provides an estimate of the mean annual growth rate (m.a.g.r.) for the data time period and a regression coefficient, which is indicative of the extent to which the data fit the selected relationship. The closer the regression coefficient is to 1, the greater the reliability of the relationship is. Details of the use of regression coefficients are given in all textbooks on statistics.

For a data set of N data pairs $x(1),y(1)$; $x(2),y(2)$; . . . ; $x(N),y(N)$, linear regression analysis results in the equation

$$\hat{y} = a + bx \tag{2.12}$$

where $\hat{y}$ = the expectation value of y for a given value of x
     a = the intercept of y for $x = 0$
     b = the slope of the linear relationship

Exponential regression analysis results in the equation

$$y = a\ e^{bx} \qquad (2.13)$$

or, in linear form,

$$\ln(y) = \ln(a) + bx \qquad (2.14)$$

where $\ln(a)$ = the intercept, from which $a = e^{\ln(a)}$
b = the mean annual growth rate (m.a.g.r.) of y if x is a time series in years

An example of an exponential regression is shown in Table 2-7 for world population from 1980 to 2000 with data obtained from the U.S. Census Bureau at www.census.gov/ipc.

Table 2-7   Regression analysis of world population 1980–2000

	Population				
Year	Billions	Year	Billions	Year	Billions
1980	4.457	1980	4.457		
1981	4.533	1981	4.533		
1982	4.613	1982	4.613		
1983	4.693	1983	4.693		
1984	4.774	1984	4.774		
1985	4.855	1985	4.855		
1986	4.937	1986	4.937		
1987	5.024	1987	5.024		
1988	5.110	1988	5.110		
1989	5.196	1989	5.196		
1990	5.284	1990	5.284	1990	5.284
1991	5.367			1991	5.367
1992	5.450			1992	5.450
1993	5.531			1993	5.531
1994	5.611			1994	5.611
1995	5.691			1995	5.691
1996	5.769			1996	5.769
1997	5.847			1997	5.847
1998	5.925			1998	5.925
1999	6.002			1999	6.002
2000	6.080			2000	6.080
m.a.g.r. (%/a)	1.57		1.70		1.40

In spreadsheet format, a comparison can be made of the mean annual growth rate during the two decade periods. The mean annual growth rate for the 1980–2000 period of 1.57%/a (percent per annum) falls between the value of 1.70%/a for the period 1980–1990 and 1.40%/a for the period 1990–2000. Which m.a.g.r. value should be used to extrapolate these data to forecast the world population in, say, 2010 or 2020? The results of the regression analysis suggest that the growth rate is falling per decade. Thus, a judgment must be made whether the growth rate will continue to fall during the next one or two decades or whether this variation in 10-year mean annual growth rates is within the normal standard deviation of 20-year population growth rates. Models in the field of mathematical statistics can help in making this decision, but the use of either value should allow a forecast to be made (albeit a more uncertain one).

2.3   CURRENT GROWTH IN ENERGY CONSUMPTION

The year 1800 may be taken as the beginning of the current period in terms of the rapid growth of energy consumption. Figure 2-8 shows the rapid growth of fossil fuel and hydropower energy consumption in the United States (from Hubbert [2]) from 1800 to the 1960s and the abrupt change in the mean annual growth rate after the turn of the century (Hubbert [2]).

2.31   Trends in Energy Consumption

There was an abrupt change in the growth rate of U.S. energy consumption after 1900 from a mean of 7.4%/a to a mean of 2.04%/a by 1960, during which time energy consumption rose from 15 quads/year in 1907 to 45 quads/year in 1960. These data can be compared to U.S. Department of Energy data for the last century (Table 2-8). This decline in growth rate indicates the nonsustainability of high growth rates in the demand for fossil fuels.

The steady decline in growth rate to 1.7%/a by 2000 in the United States can be compared with the world mean value over the same century of 1.4%/a for energy consumption, which increased from 348 quads in 1900 to 399 quads in 2000.

The trend in energy consumption was influenced greatly by the changes in energy type over the century. The major energy source in

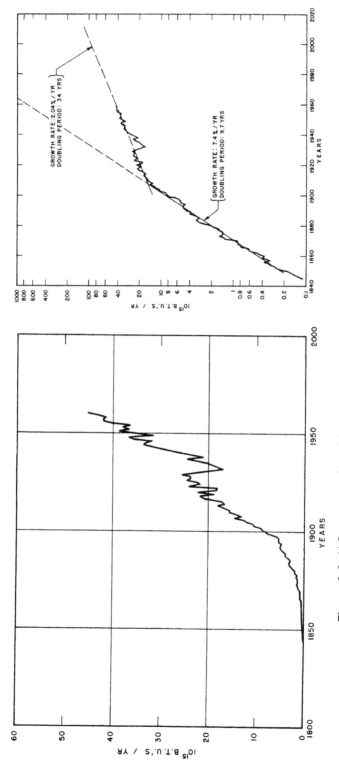

Figure 2-8. U.S. consumption of fossil fuel and hydropower energy 1800–1960s and the change in mean growth rate after the turn of the century [2].

Table 2-8   U.S. energy consumption 1900–2000

Year	Consumption (Quads/yr)	m.a.g.r. (%/a)
1900	5	—
1960	45	3.7
1990	100	2.7
2000	119	1.7

the United States in 1900 was coal, which in 1920 accounted for 89% of total energy consumption, but after the 1920s, coal was replaced significantly by oil and natural gas. Hubbert [2] notes that by 1960, coal energy consumption had dropped to 23%, whereas oil and gas energy consumption increased from 8% in 1920 to 73% in 1960.

Another major trend in the last century, especially in the United States, was the rapid growth in electric energy generation, which rose from 0.4 PWh in 1950 to 3.9 PWh in 2000. A petawatt-hour (PWh) is 10^{15} watt-hours (Wh). Here again, a decrease in the mean annual growth rate over the second half of the century is observed, as can be seen from the data in Table 2-9.

2.32   Energy Intensity

The distribution of energy consumption can be examined in terms of energy intensity, which describes the amount of energy used by a population or nation in units of energy per capita or energy per monetary unit of gross domestic product (GDP). These statistics are compiled and reported annually [3] by the U.S. Department of Energy. A summary of the data for 1990 and 2000 is given in Table 2-10 from the 2003 report [3] for both the United States and the world.

Table 2-9   U.S. electric energy generation 1950–2000

Period	Generation (PWh/yr)	m.a.g.r. (%/a)
1950–2000	0.4–3.9	4.8
1980–2000	2.2–3.9	2.8
1990–2000	2.9–3.9	2.4

Table 2-10 Energy intensities for the United States and the world

Year	Energy (Quad)		GDP (T'97$)		Population (Billions)		Intensity (kBtu/$)		Intensity (MBtu/cap)	
	United States	World	United States	World	United States	World	United States	World	United States	World
1990	101	348	6.84	24.4	0.255	5.255	14.7	14.3	395	66.3
2000	119	399	9.37	31.9	0.276	6.049	12.7	12.5	430	65.9
m.a.g.r. (%/a)	1.65	1.35	3.14	2.67	0.79	1.40	−1.50	−1.32	0.86	−0.06

2.33   Projections of Energy Intensities

The U.S. Department of Energy also provides projections of energy intensities, currently through the year 2025. The current projections [3] are summarized in Table 2-11. These projections follow from a large number of assumptions considered by the staff of the Energy Information Agency (EIA) of the Department of Energy. Several observations can be made from a comparison of the projections in Table 2-11 with the historic data in Table 2-10. One of these observations is the relative equality between the United States and the world in terms of energy intensity based on gross domestic product, whereas the disparity in per capita energy consumption between the United States and the world is sixfold. A second observation is the steady decline over the 125-year period in energy intensity based on gross domestic product compared with the very slow increase (m.a.g.r. $<1\%$/a) in energy intensity based on population. These data merit careful consideration in looking ahead for the next 50 years.

2.34   Projections of Future Primary Energy Consumption

Before examining the history of and expectations for the fossil fuel age in the next chapter, it will be instructive to review the projections of the Energy Information Agency of the U.S. Department of Energy on the distribution of primary energy sources that are expected to be used in the United States over the next two decades. It is also instructive to compare the projections prepared through 2020 in the 1999 report [4] with the projections prepared for 2020 and 2025 in the 2004 report [5]. The data from these two report are given in Table 2-12. The mean annual growth rates are given in %/a from 2000.

The data show several distinct trends in the EIA's expectations for the several resources of primary energy available in the United States. The first three, the major fossil fuels, grow from 85.6% of the total in 2000 to 87.5% in 2025. The data for nuclear energy show the reluctance of the agency to accept the need for nuclear power in the future, increasing the expectation because of increased efficiency in operation and the renewal of licences for many of the existing nuclear power plants from 40 to 60 years, but with the expectation that no nuclear power plants will be built in the United States over the next 20 years. The data for renewable energy are mostly for existing hydropower stations with little expectation that additional dams will be installed and assuming that solar, wind, and biomass energy will grow very slowly.

Table 2-11 Projection of energy intensities for the United States and the world

Year	Energy (Quad)		GDP (T'97$)		Population (Billions)		Intensity (kBtu/$)		Intensity (MBtu/cap)	
	United States	World	United States	World	United States	World	United States	World	United States	World
2005	125	433	10.6	36.3	0.288	6.43	11.8	12.0	433	67.4
2015	149	532	14.6	50.1	0.313	7.20	10.2	10.6	475	73.9
2025	171	640	19.3	67.4	0.338	7.93	8.9	9.5	507	80.7
m.a.g.r. (%/a)	1.7	1.9	3.0	3.1	0.8	1.1	−1.4	−1.2	0.8	0.9

Table 2-12 Projections of future primary energy consumption in the United States

Resource	Actual 2000		Forecast [5] 2020			Forecast [6] 2020			Forecast [6] 2025		
	Quads	%	Quads	%	m.a.g.r.	Quads	%	m.a.g.r.	Quads	%	m.a.g.r.
Petroleum	38.5	38.7	48.1	40.1	1.11	52.6	40.4	1.56	56.6	40.7	1.53
Natural gas	24.1	24.2	33.2	27.7	1.60	33.0	25.4	1.57	35.8	25.7	1.58
Coal	22.6	22.7	26.3	21.9	1.74	27.7	21.3	1.00	29.4	21.1	1.04
Uranium	7.9	7.9	3.8	3.2	−0.44	8.4	6.5	0.34	8.4	6.0	0.27
Renewables	6.0	6.0	8.2	6.8	0.67	8.3	6.4	1.64	8.8	6.3	1.55
Other	0.3	0.3	0.4	0.3	1.15	0.2	0.2	−3.0	0.1	0.1	−6.0
Total	99.4		120.0		0.94	130.1		1.34	139.1		1.34

Other primary sources include geothermal energy. The potential for these resources will be examined in subsequent chapters.

2.4   SUMMARY

This chapter continued the discussion of the human quest for abundant energy with a review of growth in energy demand from prehistoric times to today and with forecasts into the near future. Emphasis was placed on the growth of energy consumption after the Industrial Revolution and through the current nuclear and electronic eras. Accelerated growth resulted during this period from technical advances by industrial nations that substituted dependence on the energy flow from the sun to dependence on manufactured fuels from fossil combustible materials extracted from the earth's crust, a change from income (everyday solar energy) resources to capital (exhaustible fossil fuels), as postulated as Axiom 2. A brief review of the mathematics of exponential growth was included to assist with interpretation of the data available on the growth rates of energy consumption today and into the future. The chapter concluded with a review of trends in energy consumption in preparation for examining the history and future of available energy resources for sustained growth of the energy supply for the growing population of the United States and the world, along with an examination of the effects of the resulting environmental impacts on both global and local scales.

REFERENCES

[1] E. Cook, "Energy Flow in an Industrial Society." *Scientific American* 225(3): 1971.

[2] M. King Hubbert, *Energy Resources: A Report to the Committee on Natural Resources.* National Academy of Sciences–National Research Council Publication 1000-D. Washington, DC, 1962.

[3] U.S. Department of Energy, Energy Information Agency, *International Energy Outlook.* Report No. DOE/EIA-0484(2003). Washington, DC: U.S. Department of Energy, 2004.

[4] U.S. Department of Energy, Energy Information Agency, *Annual Energy Outlook.* Report No. DOE/EIA-0383(1999). Washington, DC: U.S. Department of Energy, 2000.

[5] U.S. Department of Energy, Energy Information Agency, *Annual Energy Outlook*. Report No. DOE/EIA-0383(2004). Washington, DC: U.S. Department of Energy, 2004.

3

THE FOSSIL FUEL ERA

3.0 HISTORICAL PERSPECTIVE

It may be said that the fossil fuel era began when fire was discovered in prehistoric times. More often, it is said that the fossil fuel era began with the advent of the Industrial Revolution in the 1700s. Although wood was the first combustible fuel for thermal energy after fire was discovered and remained in use for many thousands of years, the use of a higher-specific-energy fuel became desirable by the thirteenth century for production of charcoal. The use of coal, the first fossil fuel employed in quantity, increased steadily over the next 400 years. The picture of the world use of fossil fuels since 1800 is well documented in the literature. Figure 3-1 illustrates the history of fossil fuel consumption by fuel type [1] from 1800 to 2000 accented by the decline of traditional (renewable) biomass (essentially wood).

The data show that coal consumption passed wood consumption before 1900 and peaked in the early 1900s as the demand for oil and natural gas increased rapidly. The rise in commercial nonfossil fuel after 1920 was due primarily to hydropower, and the further increase since the 1970s indicates the beginning of the nuclear energy era. Data [1] for the three fossil fuels from 1765 to 1995 [totaling 12, 200 exajoules (EJ)] are reproduced below. It has been noted (see Chapter 5) that the corresponding release of CO_2 to the atmosphere accounted for 250 gigatons of carbon (GtC).

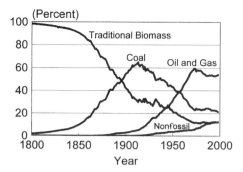

Figure 3-1. Fractional share of world consumption by fossil fuel type 1800–2000 [1] (adapted from S. Fetter data, 1999, and private communication, 2005.)

Fossil Fuel	Consumption (EJ)
Coal	5300
Oil	4800
Gas	2100

3.01 Fossil Fuel Consumption in the United States since 1900

The history of the consumption of fossil fuels in the United States follows a pattern similar to that of the rest of the world. Figure 3-2 shows the distribution (in percent) of the consumption of fossil fuels from 1900 to 1960 relative to hydropower as given by Hubbert [2]. Coal consumption decreased in that period from 88.9% in 1990 to 23.1% in 1960, whereas oil and gas increased from 8% in 1900 to 73% in 1960. During that period, hydropower increased from 3.1% to 3.9%.

The history of the primary fuel mix in the United States from 1925 to 2000 was gathered from the literature, especially from U.S. Department of Energy/Energy Information Agency (DOE/EIA) sources on the Internet, and compiled (in percent) for fossil, nuclear, and renewable resources in 25-year periods. Table 3-1 provides a summary of the data, showing many significant changes during those 75 years.

Recent data for the production of fossil fuels in the United States were compiled by the DOE/EIA [3] since 1987 and are shown in Figure 3-3. The units for coal are billion short tons; for crude oil, billion U.S. barrels, and for natural gas, trillion cubic feet.

Figure 3-3 also shows the mean annual growth rates over that period. There were steady increases for coal and natural gas (both used for increased electric power generation) and a dramatic decrease for crude

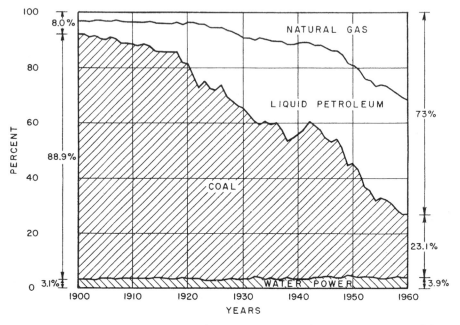

Figure 3-2. Fractional share of U.S. consumption by fuel type 1900–1960 [2].

oil (symptomatic of the increasing proportion of imported oil used for transportation).

3.1   FOSSIL FUELS

The term *fossil fuels* implies that those fuels are very old. It is well understood worldwide that fossil fuels are a nonrenewable energy re-

Table 3-1   U.S. primary fuel mix history 1925–2000 (in %)

Fuel	1925	1950	1975	2000
Wood	7	3	0	0
Coal	65	37	21	32
Oil	19	39	38	21
Natural gas	5	18	32	27
Fossil	97	97	91	81
Nuclear	0	0	6	11
Renewable	3	3	3	8
Percent for electricity	<1	~2	~15	38

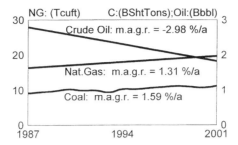

Figure 3-3. U.S. fossil fuel production 1987–2001 [3].

source. A formal definition might be as follows: Fossil fuels are the result of anaerobic decay of geologically deposited vegetation that underwent metamorphism over geologic time under lithostatic pressure and temperature. The more important fossil fuels in large-scale industrial applications of their thermal combustion energy are coal, crude oil, and, most recently, natural gas. A short review of the basic characteristics of these fossil fuels follows.

3.11 Coal

Coal, which may be described as a mineral consisting mostly of elemental carbon, exists on each continent of the world. The largest resources are found in 37 of the 50 United States ($\sim$23%), the former Soviet Union ($\sim$23%), and China ($\sim$11%). China is the world's leading producer, followed by the United States. Coal is used primarily for steam-turbine electricity generation and for coking in the iron and steel industry.

Coal exists in a large variety of geologic forms and levels of quality and, more important, specific heating value. Coal can be characterized in a continuous range of ranks, beginning with peat, then lignites (brown coal), then bituminous (soft coal), and finally anthracite (hard coal) and graphite. U.S. coal classifications by composition and heating value are listed in Table 3-2.

3.12 Heating Value of Coal

One of the key properties of coal is its heating value: the amount of energy released from its combustion as a fuel. Since most of the combustion energy comes from the oxidation of carbon, carbon is the key component. However, since the elements hydrogen and sulphur in coal

Table 3-2   U.S. classification of coal

	Composition (%)				Heating Value (Btu/lb)
	Fixed		Volatile		
Coal Type	Carbon	Moisture	Matter	Ash	
Anthracite	81.8	4.4	4.8	9.0	13,130
Bituminous					
Low-volatile	65.8	2.3	19.6	12.3	13,220
Medium-volatile	63.6	3.1	23.4	9.9	13,530
High-volatile	46.5	5.9	43.8	3.8	13,150
Subbituminous	41.0	13.9	34.2	10.9	10,330
	38.4	25.8	31.1	4.7	8,580
Lignite	30.2	36.8	27.8	5.2	6,950

have high combustion energy, they add significantly to the heating value. The heating value of the several classifications of coal results from the heat of combustion, ΔH_c, given here in Btu/lb for the oxidation reactions of the three major elements:

$$C + O_2 \Rightarrow CO_2 \qquad \Delta H_c = 14{,}500 \text{ Btu/lb}$$
$$H_2 + \tfrac{1}{2}O_2 \Rightarrow H_2O \qquad\qquad\quad 62{,}000$$
$$S + O_2 \Rightarrow SO_2 \qquad\qquad\qquad\;\; 4{,}000$$

Thus, the heating value of a coal with measured percentage concentration of the three elements is

$$\text{HV(coal)} = (14500[C] + 62000[H] + 4000[S])/100 \qquad (3.1)$$

3.13   Crude Oil

Crude oil is pumped out of the ground as petroleum, a mixture of aliphatic hydrocarbons, mainly alkanes of formula C_nH_{2n+2}, and several impurities, with sulphur being the most important. Oil with a low sulphur content is considered "sweet" crude; oil with high sulphur content is called "sour" crude. The hydrocarbons are classified by their chain lengths. Those with longer chain lengths generally have lower boiling points, which allows for their separation by distillation in refineries. They are classified by C_n content as follows:

Gases	C_1 to C_4	The most abundant is methane (CH_4, the main constituent of natural gas)
Naphthas	C_5 to C_7	Used as solvents (for quick-drying products)
Gasoline	C_8 to C_{11}	C_8H_{18} is octane, the basis for the octane rating
Kerosene	C_{12} to C_{15}	Followed by diesel fuel and heavier fuel oils
Lubricating oils	C_{16} to C_{19}	These oils vaporize at elevated temperatures
Solids	C_{20} up	These are solids at room temperature

The lubricating oils are important for transportation; engine oil can run all day at 250°F (120°C) without vaporizing. These oils vary from light (e.g., 3-in-1 Oil) through various densities of motor oil to very thick gear oils and semisolid greases such as petroleum jelly. The solid oils increase in carbon number from paraffin, to wax, to tar, and finally to asphaltic bitumen, which is used in asphalt roads.

3.14 Natural Gas

Natural gas (primarily methane, CH_4) is found in sedimentary rocks, generally in the presence of crude oil as "wet" gas (associated gas) or separately as "dry" gas (nonassociated gas). Wet gas generally contains many other hydrocarbon gases and several impurity gases that are removed before delivery. Dry gas generally does not contain many other gases besides methane. Natural gas is classified as "sour gas" if the concentration of H_2S is above a maximum permissible concentration. The composition of natural gas varies by type and deposit; typical constituents are listed in Table 3-3.

Table 3-3 Composition of natural gas

Component	Composition Range (%)
Methane	70–90
Ethane, propane, butane	0–20
Hydrogen sulphide	0–5
CO_2, O_2, N_2	0–2
Rare gases: He, Ar, Kr	Trace

The heating value of natural gas also depends on the chemical composition. The combustion energy of pure methane by

$$CH_4 + 2O_2 \Rightarrow CO_2 + 2\ H_2O + 891\ kJ \qquad (3.2)$$

is 55,700 kJ/kg. The average heating value of dry natural gas is 1027 Btu/CF.

3.2 FORECAST OF U.S. ENERGY CONSUMPTION THROUGH 2025

Projections of the demand for energy over the next 20 to 25 years are included in the annual reports of the U.S. Department of Energy [3,4]. These projections (and those of other government-supported agencies) indicate the most likely trends in the production and consumption of energy in the United States and worldwide. The 2004 report for energy consumption in the United States [3], which contains the actual values (in quads) from 1990 through 2001 and the forecast values through 2025, is illustrated in Figure 3-4.

The most likely case anticipated by DOE/EIA is steady exponential growth at a m.a.g.r. of 1.7%/a (see Table 2-11), implying the agency's belief that public pressure for increasing use of nonfossil fuels will not be implemented significantly in the next 20 to 25 years. The projections for fossil fuel consumption in the United States in comparison to the world [4] are shown in Figures 3-5 to 3-7 for coal (in billion short tons), oil (in million barrels per day), and natural gas (in trillion cubic feet), respectively.

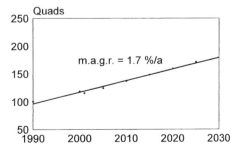

Figure 3-4. Historical and projected U.S. energy consumption 1990–2025 [3].

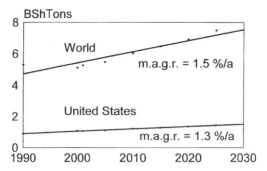

Figure 3-5. Projected consumption of coal in the United States and the world through 2025 [4].

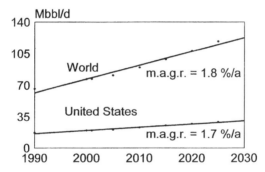

Figure 3-6. Projected consumption of oil in the United States and the world through 2025 [4].

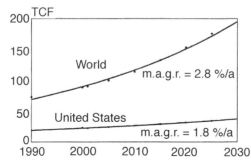

Figure 3-7. Projected consumption of natural gas in the United States and the world through 2025 [4].

3.3 HOW LONG WILL FOSSIL FUELS LAST?

Projections for energy consumption in the United States and the world through 2025 are summarized in Table 3-4.

The projections indicate a continuously increasing world dependence on the three major fossil fuels (with mean annual growth rates ranging from 1.5 to 2.8%/a). As the less-developed countries become more industrialized, there is likely to be a concomitant increased demand for higher-specific-energy fuels. This raises the questions, How much extractable fossil fuel is actually available on earth? and How long can the world depend on having sufficient fossil fuels at affordable prices?

The question about ultimate reserves of fossil fuel resources results from the long-known fact that fossil fuels (e.g., coal, petroleum, and natural gas) have taken millions of years to deposit in geologic resources. As the extraction, processing, and utilization of each of those energy sources, in turn, grew exponentially, it was believed that those energy sources would last forever or at least for a very long time. By the 1950s, concern began to build rapidly that these energy sources were finite resources and that the rapid exponential growth in their consumption would result in their early exhaustion. In 1962, Hubbert [2] prepared a picture (see Figure 3-8) of what the fossil fuel era would look like in geologic time.

3.31 Estimation of Fossil Fuel Reserves

One of the most difficult problems for the mineral extraction industries is determining how much of an exhaustible (depletable) natural resource is available in the ground so that the lifetime of that resource can be estimated. It is very difficult to estimate the reserves in an existing mineral deposit and even more difficult to estimate the reserves

Table 3-4   Projected U.S. and world energy consumption 2000–2025 [4]

Energy Source	United States			World		
	2000	2025	m.a.g.r.	2000	2025	m.a.g.r.
Coal (BShTon)	1.08	1.44	1.3	5.12	7.48	1.5
Crude oil (Mbbl/d)	19.7	29.2	1.7	76.9	119	1.8
Natural gas (TCF)	23.5	34.9	1.8	88.7	176	2.8
Nuclear (PWh)	0.75	0.81	0.2	2.43	2.74	0.3
Renewable (quad)	6.4	8.9	2.0	32.8	50.0	1.9
Total (quad)	119	171	1.7	400	640	1.9

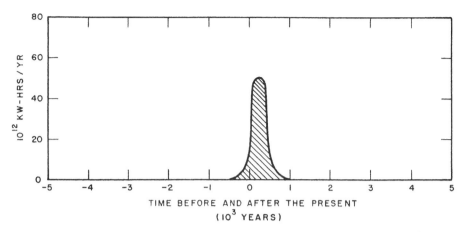

Figure 3-8. The fossil fuel era in geologic time as depicted by Hubbert [2].

of a nation or the world. Yet the financial aspects of the mineral extraction industry require this estimation for pricing, taxes, and consumer satisfaction, among other needs. Estimates generally are made by geologists on the basis of various physical methods and statistical models developed by mining and petroleum engineers. Early attempts to estimate resources in place were based on volumetric considerations of surface measurements and any drilling data available. By the 1950s, the technique of calculating reserves by using a mathematical model was based on an S-shaped logistic curve (as noted in Hubbert [2]) for cumulative production and a bell-shaped curve for the production rate.

3.32 The McKelvey Diagram

An early method to describe the potential of a mineral deposit that now is known as the McKelvey diagram was developed by the United States Geologic Survey. Since 1972, it has helped establish the vocabulary of reserves with definitions of three levels of recoverability—recoverable, paramarginal, and submarginal—for both identified and undiscovered resources. The concept was based on

1. The resource base, a geologic estimate of the total quantity of a resource in the ground that is extractable by some technology at any cost
2. Reserves, a specific estimate of recoverable resources from identified (discovered) reservoirs using current technology and at current prices

Thus, for a given mineral reservoir, the reserves could be estimated as the product of the estimated resources base and a recovery factor determined as a function of technology and price. This concept is illustrated in Figure 3-9.

The method has been extended for commercial applications to estimate reserves under more standard economic conditions to serve as the basis for financial valuations and taxes. The figure includes more boxes relating a decreasing degree of geologic certainty to a decreasing degree of economic viability.

The development of the McKelvey diagram methodology has resulted in a worldwide effort, the United Nations Framework Classification (UNFC) for Energy and Mineral Resources, with the goal of a global framework for estimating the reserves as part of the total resource. An explanation of the framework classification system for oil and gas fluids and solid minerals was presented by Ahlbrandt et al. [5]. A final report [6] to the United Nations Economic Commission for Europe (UNECE) provides details of the UNFC system.

The procedures are built around three parameters that best define the essential characteristics of a resource in market economics. The construction of the model is illustrated in Figure 3-10.

The total initial resource (in place, as in the McKelvey diagram) is accounted for with three quantities (Figure 3-10a):

1. Amount already produced
2. Estimated remaining amount recoverable
3. Possible additional amount in place

Since the produced quantity is essentially known and the additional amount is essentially unknown, the focus of the method is on the re-

	Identified	Undiscovered
Recoverable	Reserves	requires further exploration and R&D
Paramarginal	recoverable at prices to 1.5 × current	also requires economic incentive
Submarginal	recoverable with advances in extraction technology	area of wild guess

Figure 3-9. The McKelvey diagram for estimating reserves of mineral resources.

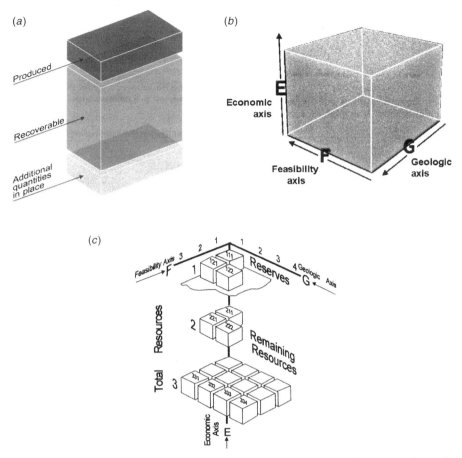

Figure 3-10. The basic cubic structure of the United Nations Framework Classification EFG system of resource evaluation [6].

maining recoverable quantity. The basis of the classification is the estimation of the total remaining resource by three criteria—the EFG system—in which

E = economic and commercial viability
F = field project status and feasibility
G = geological knowledge

The model entails detailing available information about these three criteria in the cubic format illustrated in Figure 3-10b.

An example of the three-digit codification is illustrated in Figure 3-10c for solid materials such as coal or uranium ores, where the cube labeled 111 is of primary interest. The classification for reserves implies that the quantity estimates for that cube are economically and commercially recoverable in vector E1, justified to be technically recoverable (by a feasibility study or actual production) in vector F1, and based on assured geology in vector G1. Estimates of remaining resources are expressed in subcategories of the three vectors.

The method most widely used for estimating large-scale reserves is based on modeling as a logistic curve, augmented by both production and discovery rate data. The model uses the historical annual production data cumulated over the total period of extraction and extrapolates the data, using the S-shaped logistics curve, to the ultimate extractable amount of the resource and estimates the production rate history for a finite resource as a bell-shaped curve. The area under the curve is equal to the ultimate extractable amount. The logistic curve is calculated in three time periods:

1. Early: as exponential growth to time t (current year data)
2. Midtime: estimating the inflection point at maximum extraction rate at time t_m
3. Late: as total extraction asymptotically approaches the ultimate extractable amount

This was the method used by Hubbert [2] in estimating fossil fuel resources in the United States and the world. In practice today, two additional annual and cumulative statistics are used:

1. Additional discovered resources
2. Revised estimate of remaining proved reserves.

3.33 Production of a Finite Resource

Early production statistics for a newly successful commercial primary resource follow a rapidly growing exponential curve, with production rate (P) given by

$$P = \frac{dQ}{dt} \tag{3.3}$$

where dQ is the amount produced per unit time (e.g., per year), as shown in Figure 3-11.

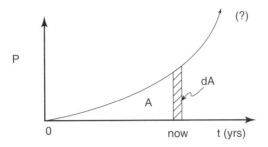

Figure 3-11. Early production rate history of a new resource.

Total production from t = 0 to "now," Q(t), is represented by the area A under the curve extending to "now." The incremental area dA is the most recent production in time dt. The early production data foster the expectation that the resource will last forever. For a finite resource, however, the total production will always be less than (or eventually equal to) the total resource, Q^∞. Thus, it easy to assume that the final production curve will resemble a bell-shaped distribution with the form shown in Figure 3-12.

Here R is the proved reserves, P(max) is the maximum production rate at time t_m in the future, and Q^∞ is the ultimate extracted resource, given by

$$Q^\infty = \int_0^\infty P \, dt \qquad (3.4)$$

3.34   The Logistic Production Curve Method

The logistic production curve method uses the two additional features noted in Section 3.32: the discovered resources curve and the revised

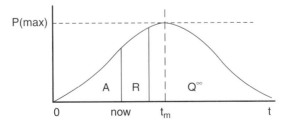

Figure 3-12. Total production rate curve for a fully extracted finite resource.

estimate of remaining proved reserves. The nomenclature for the method is listed below:

$Q_P(t)$ = cumulative production to time t

$Q_D(t)$ = cumulative discovered reserves

$Q_R(t) = Q_D(t) - Q_P(t)$ = remaining proved reserves

Δt = time lag from discovery of new resources to production

then $Q^\infty = Q_D(\infty) - Q_p(\infty)$ = ultimate recoverable reserves

and for $R = Q_D(t) - Q_P(t)$ (remaining proved reserves)

and $P = dQ_P/dt$ = annual production rate

R/P = lifetime of the resource at constant current production rate.

But. . . what is the value of Q^∞?

The cumulative production curve (which comes from the integral of the production curve shown in Figure 3.12) follows the logistic growth curve (see Section 2.24) by Eq. 2.11. The relationship between cumulative discoveries, cumulative production, and proved reserves is shown in Figure 3.13.

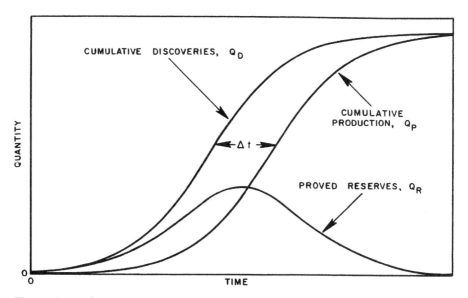

Figure 3-13. Cumulative discoveries, cumulative production, and proved reserves [2].

The lag time, Δt, which varies with time, is estimated by matching the two logistic curves, and the best value of the mean Δt near the inflection points of the two logistics curves is used to obtain the values of a, b, and Q^∞ from the two equations:

$$Q_D = \frac{Q^\infty}{1 + a\,e^{-bt}} \quad \text{and} \quad Q_P = \frac{Q^\infty}{1 + a\,e^{-(bt+\Delta t)}} \qquad (3.5)$$

An example of the method that was used by Hubbert [2] to estimate the ultimate recoverable reserves of crude oil in the United States is shown in Figure 3-14.

A summary of the results of the analysis of Hubbert [2] for the fossil fuel resources of the United States and the world for data though the

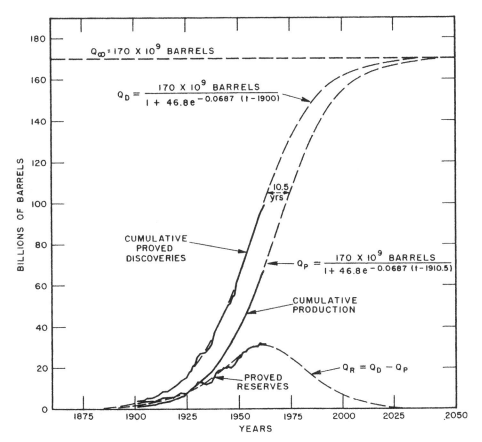

Figure 3-14. Logistic curve method used for U.S. crude oil production data by Hubbert [2].

1960s is listed in Table 3-5. The units used for the estimated ultimate reserves are billion metric tons (BMt) of coal, billion barrels (Bbbl) of oil, and trillion cubic feet (TCF) of natural gas, and the units used for the thermal energy content are exawatt-hours (EWh = 10^{18} Wh).

Another general tabulation of world fossil fuel resources was published by Cassedy and Grossman [7], using literature data compiled from 1979 to 1995. Table 3-6 lists the data rearranged to show the latest production rate, P; remaining reserves, R; current longevity, R/P; ultimate resource, Q^{∞}; and remaining undiscovered resources, $(Q^{\infty} - Q_D)$.

More recent estimates of U.S. and world fossil fuel reserves [8], compiled from reports of the World Energy Council, *BP Statistical Review of World Energy,* and DOE/EIA, are listed in Table 3-7. The initial year for coal is 1997 instead of 1992.

It is interesting to note the changes in estimates of world fossil fuels over the 40-year period of continuous development of more sensitive technical methods and models for estimating the total resources, remaining reserves, and useful lifetimes of the finite fossil fuel resources of energy. The values for coal decreased steadily from 2400 BMt (1962) to 1038 BMt (1995) to 984 BMt (2002). The corresponding values for crude oil were 1250, 1103, and 1047 Bbbl. The estimation of natural gas reserves is still difficult because of the lack of extensive production and flaring data. The subject of natural gas utilization will be covered in Chapter 4, which will examine the sustainability of natural gas to satisfy the existing and new demands for natural gas as a chemical feedstock, for heating and cooling of commercial and residential buildings, and for use as an energy source for electricity generation and transportation fuel.

Table 3-5 Fossil fuel ultimate reserves and energy content estimated in the 1960s [2]

Fossil Fuel	United States		World	
	Reserves	Energy	Reserves	Energy
Coal (BMt)	875	6.8	2400	19.6
Crude oil (Bbbl)	170	0.25	1250	2.4
Natural gas (TCF)	~1000	0.25	~7500	2.4
Shale oil (Bbbl)	850	1.4	1300	2.2
Tar sands (Bbbl)	2.6	0.0	>490	0.8
Total thermal energy (EWh)		8.7		27.4

Table 3-6 Estimated world fossil fuel resources 1979–1995 [7]

	Coal			Crude Oil			Natural Gas		
	BMt/a	BMt	Years	BMt/a	BMt	Years	BMt/a	BMt	Years
Production rate (P)	3.53			22			73		
Reserves (Q_R)		1038			1103			4980	
Longevity (R/P)			294			50			68
Resource (Q^∞)		~5380			2273			11,561	
Remaining ($Q^\infty - Q_D$)		~5400			471			4687	

Table 3-7 Current estimated world fossil fuel resources [8]

	(Year)	Coal		Crude Oil		Natural Gas	
		Reserves (BMt)	Longevity (R/P, years)	Reserves (Bbbl)	Longevity (R/P, years)	Reserves (TCF)	Longevity (R/P, years)
United States	1992	241	244	32.1	10	167	9.4
	2002	250	252	30.4	11	184	9.6
World	1992	1031	219	1007	42	4890	66
	2002	984	204	1047	41	5500	61

3.4 GROWTH OF FOSSIL FUEL DEMAND FOR GENERATION OF ELECTRICITY

Electric energy has become the predominant energy form in the industrialization of the United States. The growth of electricity generation since 1950 was noted in Table 2-9 in Section 2.31. The growth over this period has been fueled primarily by fossil fuels. The data on electricity generation by fuel from the DOE [3] are summarized by fuel in Table 3-8 and by fuel type in Table 3-9. Figure 3-15 shows the distribution of fossil, nuclear, and renewable fuels over this period. The fraction of electricity generated from fossil fuels ranged from 69% to 80% during this period. In the last decade, from 1990 to 2000, electricity generation grew from 3.0 to 3.8 PWh/yr at a mean growth rate of 2.36%/a.

Several observations may be derived from these data. First is the steady growth in electrification of the United States from 1950 at a mean annual growth rate of 4.78%/a (a doubling time of 15 years!). Second is the dominant fraction (70 to 80%) accounted for by fossil fuels. Fossil fuel use for electricity generation showed two significant changes. Coal was the primary fuel for generating electricity, growing steadily at an m.a.g.r. of almost 5%/a, whereas the oil fraction increased only until 1980 and then deceased when oil was used primarily for the transportation sector. Almost all renewable energy used during this period came from hydroelectric power. The decrease in the fossil fuel fraction since 1980 was offset by the rapid growth of nuclear power from the 1950s until about the 1980s, when antinuclear sentiments resulted in no further construction of nuclear power plants. That fraction continued to increase somewhat as the efficiency of existing nuclear power plants increased markedly.

Table 3-8 Electricity generation (PWh) in the United States by fuel 1950–2000 [3]

Year	Coal	Oil	Natural Gas	Uranium	Hydropower	Other
1950	0.155	0.034	0.045	0.000	0.101	0.000
1960	0.403	0.048	0.158	0.001	0.149	0.001
1970	0.704	0.184	0.373	0.022	0.251	0.001
1980	1.162	0.246	0.346	0.251	0.279	0.006
1990	1.594	0.127	0.373	0.577	0.293	0.064
2000	1.966	0.125	0.601	0.754	0.276	0.081
m.a.g.r (%/a)	4.95	2.78	4.42	22.8	2.05	n/a

Table 3-9 Electricity generation in the United States by fuel type 1950–2000 [3]

Year	Fossil PWh	Fossil %	Nuclear PWh	Nuclear %	Renewable PWh	Renewable %	Total (PWh)
1950	0.233	70	0.000	0.0	0.101	30	0.334
1960	0.609	80	0.001	0.1	0.150	20	0.760
1970	1.262	82	0.022	1.4	0.252	16	1.536
1980	1.754	77	0.251	11.0	0.285	12	2.290
1990	2.104	69	0.577	19.0	0.357	12	3.038
2000	2.693	71	0.754	19.8	0.357	9	3.804
m.a.g.r (%/a)	4.65		22.8		2.58		4.78

The 2004 forecast of electricity generation by DOE/EIA [3] (Figure 3-16) shows a continued growth, at an m.a.g.r. of 1.9%/a, to a demand of 5.252 PWh in 2025, with fossil fuels providing the bulk of the primary energy. Although coal remains the major fraction, a large acceleration in the use of natural gas is expected through 2025, Nuclear energy is maintained constant as no new power plants are expected but the licensed lifetime of many of the existing plants is extended for 20 years. The forecast also shows a further decline in the use of oil for electricity generation and a modest growth in the use of renewables other than hydropower.

3.5 SUMMARY

The chapter examined the role of fossil fuels in the supply of energy since the Industrial Revolution. It showed the continuous change from

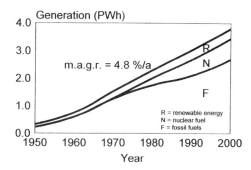

Figure 3-15. Electricity generation by fuel type in the United States 1950–2000 [3].

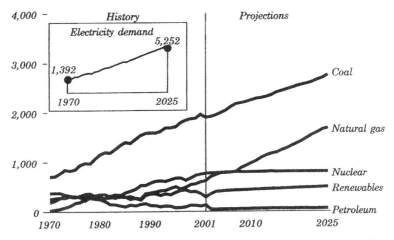

Figure 3-16. DOE/EIA forecast of electricity generation by fuel [3].

wood as the primary fuel to the periodic switch in fossil fuels, progressing in terms of increased specific energy and *comfort and ease* handling from coal to petroleum to natural gas. Data for energy consumption in the United State in the past century show a steady average growth of about 5%/a, with fossil fuels building up to about 80% of the total energy consumption. Electric energy generation grew to about 38% of primary energy consumption by the end of the century. The procedure for estimating the amount of fossil fuels available for the future and their longevity at current levels of production was examined. Growth of population and affluence worldwide probably will shorten that longevity if discoveries of new resources decline. A look ahead to the year 2025 shows a continued dependence in the United States (and the world) on continued growth in fossil fuel consumption. Chapter 4 examines the question of the sustainability of energy consumption, especially as it affects the rapidly growing demand for natural gas as a chemical commodity and an energy fuel.

REFERENCES

[1] S. Fetter, *Climate Change and the Transformation of World Energy Supply.* Center for International Security and Cooperation (CISAC) Report. Stanford, CA: Stanford University, May 1999.

[2] M. K. Hubbert, *Energy Resources: A Report to the Committee on Natural Resources.* National Academy of Sciences–National Research Council Publ. 1000-D. National Academies Press, Washington, DC, 1962.

[3] U.S. Department of Energy, Energy Information Agency, *Annual Energy Outlook 2004*. Report No. DOE/EIA-0383(04) (and earlier years). Washington, DC: U.S. Department of Energy, 2004.

[4] U.S. Department of Energy, Energy Information Agency, *International Energy Outlook 2003*. Report No. DOE/EIA-0393(03). Washington, DC: U.S Department of Energy, 2004.

[5] T. S. Ahlbrandt et al., *Updated United Nations Framework Classification for Reserves and Resources of Extractive Industries*. Paper SPE 90839, Society of Petroleum Engineers Annual Technical Conference, Houston, TX, September 26–29, 2004.

[6] United Nations Economic Commission for Europe, *Final Report on the United Nations Framework Classification (UNFC) for Energy and Mineral Resources*. Available at www.unece.org/ie/se/reserves.html, 2005.

[7] E. S. Cassedy and P. Z. Grossman, *Introduction to Energy*, 2nd ed. New York: Cambridge University Press, 1998.

[8] BP p.l.c., *BP Statistical Review of World Energy*, June 2005. Available at www.bp.com/statisticalreview2005.

4

SUSTAINABILITY OF ENERGY RESOURCES

4.0 SUSTAINABLE ECONOMIC DEVELOPMENT

The continuous growth of energy demand with growth of population and economic development, with gross domestic product (GDP) increasing at rates of 2 to 5%/a, as discussed in the last three chapters, raises the question of whether the quest for abundant energy can continue far into the future. The question becomes even more important as economic development accelerates worldwide. A key factor in examining the question is the potential for long-term sustainability of energy resources. This question has become more complex with social changes in the definition of economic development. A generation ago, the accepted definition of sustainable economic development was development that could be sustained indefinitely without periodic cycles of prosperity and recession. Today, the definition of economic development also includes the proviso that it not have an impact on our environmental and social awareness or impose a burden on future generations.

Environmental and societal awareness that takes into account the strategic importance of energy has resulted in a number of *solutions* for ensuring an adequate supply of energy that is sustainable and is under *reasonable* means of control. The solutions have varied from outright private ownership to outright state monopoly of energy supply. The results have varied markedly; in some cases, cyclic change between the two philosophies has occurred with changes of government administration by opposing political parties. In many nations today, the trend is towards a "competitive market" philosophy, with less government intervention. The *optimum* means for ensuring a sustainable energy supply in a widely diversified world has not been determined.

The complexity of this quest for abundant energy results from the large number of factors that must be considered (and agreed upon). Those factors include technical problems, economic limitations, environmental concerns, and a demand for social equity. The technical problems are relatively easier to examine. They cover the aspects of energy supply, involving the fuel cycle technology for delivering sufficient energy on demand. Those aspects include determining what fuels are available in terms of resource base, proved reserves, optimal conversion processes, sustainable production rates, and increased efficiencies. The economic limitations that have been examined by the financial community bear on what is worth doing. They involve aspects of the fuel cycle infrastructure and cost-price relationships that are affected by the value of money, taxation and subsidy policies, and corporate management and governmental regulation. Environmental concern, the subject of Chapter 5, which results from the desire for a *pleasant habitat* and a life of *comfort and ease,* has led to extensive research into all magnitude and severity impacts of all human activities as well as methods of analysis for examining the relative impacts over the entire fuel cycle of each of the energy resources that is available in the energy mix. The demand for social equity has raised many questions that involve the concepts of *responsibility* and *fairness.* This factor is the most difficult to resolve.

4.01   Indicators for Sustainable Energy Development

The difficulty in determining what sustainable energy levels should be lies in the method for calculating how much of what energy mix at what growth rate constitutes the *output* for the complex mix of *demands.* Many governments and institutions have addressed these questions. A project that has investigated the means to evaluate the economic, environmental, and social aspects of the problem involves a consortium of institutions and was initiated by the International Atomic Energy Agency (IAEA) in 1999 with assistance from the International Energy Agency (IEA) of the Organization of Economic Cooperation and Development (OECD) and the United Nations. This project focused on preparing a comprehensive set of quantifiable parameters (indicators) for attaining sustainable energy development. A report by the IAEA and IEA [1] lists 41 environmental and socioeconomic indicators for energy development and a model of their interrelationships.

Some of the environmental indicators are important in regard to major environmental problems such as global climate change and urban air pollution, which are discussed in Chapter 5. For example, three of the environmental indicators are quantities of greenhouse gas emissions, quantities of air pollutant emissions, and ambient concentrations of pollutants in urban areas. Socioeconomic indicators include population and GDP, energy mix, consumption, intensities, prices, and transportation aspects of road traffic per capita

4.02   Sustainable Energy Supply

The attainment of a long-term sustainable energy supply requires an equilibrium between the industrial (private) sector and the social (government) sector of the economy. This has always been a difficult problem because of the differing fundamental objectives of the two sectors. Public expectations of industrial competition should include the stimulation of innovation, increases in productivity, and improvements in resource allocation and fuel conversion efficiency, with resulting lower prices to consumers. These expectations and results pose a dilemma in the energy sector in that greater efficiency in fuel use benefits both the economy and the environment but lower prices discourage efficiency in end use; that is, wasteful use of energy is not compatible with environmental objectives. An example of the dilemma is the global problem of greenhouse gas emission. The severity of climate change raises the fear of adverse effects on the environment. However, climate change is an externality in the business of energy supply; avoidance of adverse effects is not valued in the marketplace because the benefit of avoiding global warming comes in the form of *problems avoided* rather than *marketable commodities created.* The solution to such dilemmas might be a long-term equilibrium between competition (to reduce prices by the private sector) and regulation (to enforce conservation by the government).

4.1   SUSTAINABILITY OF ELECTRIC ENERGY DEMAND

The historical growth in demand for *clean, convenient* electric energy (see Table 2-6) in labor-saving appliances from toothbrushes to skyscraper elevators by itself would cause concern about the availability of abundant electric energy for a growing population. However, new electric-energy-intensive applications of new technologies already are becoming visible. The increase in demand could be offset partially by more stringent efforts toward technical (do better) and social (do without) conservation of electricity in a smaller sphere of applications.

If it were possible to predict the business-as-usual (B.a.U) rate of growth in demand and the growth of conservation results, the increase in electric energy demand (ΔED) could be forecast by

$$\Delta \text{ED} = \int_{\text{now}}^{\text{later}} (\text{B.a.U})_0 \, e^{gt} \, dt - \int_{\text{now}}^{\text{later}} (\text{Cons})_0 \, e^{ct} \, dt \qquad (4.1)$$

where g = business-as-usual growth rate from t_0 = now to t = later
          c = the corresponding growth rate for conservation

The business-as-usual growth rate would have to be adjusted for new demands for electricity resulting from expected growth in the *electronic way of life,* the potential for the construction of a U.S. continental superconducting grid network, and the foreseeable large increase needed for the production of hydrogen for the coming hydrogen fuel age.

4.11   The Electronic Way of Life

The electric way of life may be said to have started in the early 1900s with the development of commercial radio broadcasting, the telephone, and the motion picture industry. Rapid acceleration in demand occurred during and after World War II with the introduction of more powerful radio tubes, transistors, and klystrons. However, the demand for electric energy for these industries may appear small compared with the coming growth in new electronic industries, which include computers (information technology), mobile cell phones, home management systems, aviation security, and homeland defense. The magnitude of the long-term demand for the additional electric power required to supply electricity for these new applications may account for a significant fraction of the current business-as-usual demand and is difficult to estimate at this time.

4.12   A Continental Superconducting Grid

The electricity distribution system in the United States (including Canada and a small part of northern Mexico) was divided into a number of regional systems under the guidance of the North American Electricity Reliability Council (NERC) to ensure continuity of supply over economically viable transmission distances. A long-term hope has been the possibility of a single national grid, but large losses in electric power over long distances have made this hope uneconomical for moving electricity loads through the four time zones of the country. The idea of a continental superconducting grid has been considered for almost 40 years as a means to augment the existing grid over the length of the United States. The high cost of early superconductor materials made the idea too expensive for practical consideration. Recently, the Electric Power Research Institute (EPRI) proposed the use of lower-cost MgB_2 superconductor to make the idea feasible in this century [2].

The concept behind the proposal is to construct a coast-to-coast transmission corridor underground that contains a transmission line based on "low-cost" MgB_2 superconductor cooled by liquid hydrogen (LH_2). Power plants would be spaced along the corridor at optimum distances to generate electricity and liquid hydrogen used to cool the line to the superconducting temperature of $\sim 25°K$. A cross section of the transmission line is shown in Figure 4-1. A section of the underground transmission corridor is shown in Figure 4-2.

The advantages of the concept include (1) the ability of the grid to supplement the regional electric power grids, (2) the energy savings from load leveling across the four time zones of the United States, (3)

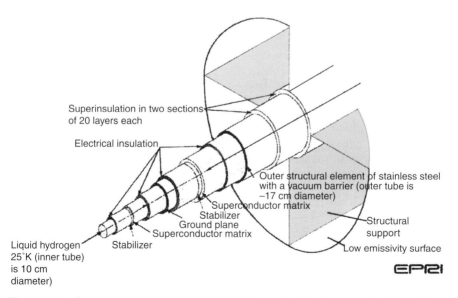

Figure 4-1. Cross section of an MgB_2 DC superconducting transmission line. (Courtesy of the Electric Power Research Institute.)

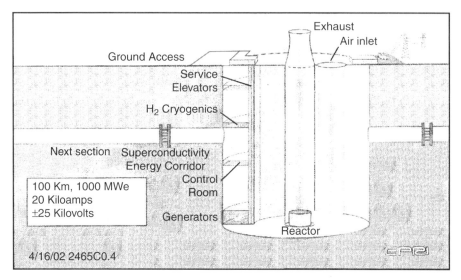

Figure 4-2. A section of the underground supergrid transmission corridor. (Courtesy of the Electric Power Research Institute.)

the possibility of combining renewable and nuclear energy as a clean, sustainable electric power supply, and (4) the synergy of sending the "spent" LH_2 coolant to the corridor surface for sale as compressed gaseous hydrogen (GH_2) fuel for hydrogen fuel-cell vehicles. A disadvantage would be the large inertial effort to initiate the new infrastructure to make it happen smoothly and economically.

4.13   The Hydrogen Fuel Era

One of the most energy-intensive changes in societal development has been the introduction of new major technologies, for example, the iron horse for the Conestoga wagon, the airplane for the iron horse, and the automobile for the horse and buggy. Each of these innovations required the introduction of a new form of energy fuel: coal for oats, aviation fuel for coal, and gasoline for oats. With the ongoing development of the fuel-cell engine that is likely to replace the internal-combustion engine over the next 50 years, the use of transportation fuel again will undergo a drastic change, from liquid fossil fuels to hydrogen. The corresponding additional demand for electricity to produce the hydrogen fuel in large enough quantities by several energy-intensive methods surely will add a large burden to the quest for abundant energy. This, as a major issue in sustainable energy growth, is the subject of the final three chapters of this book.

In summary, forecasting future energy demand, in itself a very uncertain science, is especially vulnerable to the choice of time frame. U.S. government administrations worry about the next four years. Some U.S. and state government agencies look ahead 5 to 20 years. Essentially none look ahead 30 to 50 years, the time needed for a new technology to become a normal part of everyday life. Thus, the introduction of a new fuel in a new technology is likely to have a tough time getting under way. Nevertheless, the quest for abundant energy eventually will make this happen.

4.2   NATURAL GAS IN SUSTAINABLE ENERGY SUPPLY

Natural gas (mostly as methane, CH_4) has become the fossil fuel of choice. It is relatively cheap, it burns with a nice blue hot flame, and it is easy to pipe into factories and homes, making it a desirable *comfort and ease* fuel. Methane has the largest hydrogen/carbon ratio of the alkane gases (C_xH_{2x+2}), with a value of $H/C = 4$, which offers both industrial use as a chemical raw material for producing hydrogen and the environmental benefit of a relatively low CO_2 release to the atmosphere in its combustion as a fossil fuel.

The good qualities of natural gas and its rapid growth as a fuel (see Chapter 3) raise many concerns about its sustainability for satisfying the demand for its many large-scale and competing uses. Table 4-1 lists some of the questions raised about its many desired applications,

Table 4-1   Factors that raise questions about the sustainability of natural gas demand

Competing uses for natural gas
   Chemical versus energy utilization?
Cause of rapid growth of natural gas utilization
   Growth in consumption fed by supply or demand?
Estimates of natural gas resource and reserves
   How much natural gas is available for how long?
Natural gas for electric power generation
   How will your grandchildren afford to heat their homes?
Natural gas as an energy resource
   Electricity generation and/or hydrogen fuel production?

factors that involve the sustainable economic development issues of technology choices, economic feasibility, environment protection, and social awareness.

These factors are considered in further detail below. The many uses of natural gas can be grouped into four general classes:

1. Petrochemical feedstock for chemical synthesis
2. Combustible fuel for residential and commercial heating and cooling
3. Combustible fuel for electric power generation
4. Reformer feedstock for hydrogen fuel production

4.21 Petrochemical Use of Natural Gas

Natural gas is an important feedstock component of the world's petrochemical industry that is built on the chemical engineering development of petroleum refining. A general listing of petroleum–natural gas raw material processing into primary petrochemicals (used for further chemical syntheses) is shown below. Natural gas as a petrochemical feedstock also is used for production of ammonia, acetylene, urea, and many other commercial organic products.

Raw Material Feedstock	Component	Primary Petrochemical
Natural gas		Methanol (CH_3OH)
	Ethane	Ethylene (C_2H_4)
	Propane	Propylene (C_3H_6)
Petroleum	Naphtha	Butadiene (C_4H_6)
		Benzene (C_6H_6)
		Toluene $(C_6H_5–CH_3)$
		Xylene $(C_4H_4–2CH_3)$

A key aspect in the competition for the application of natural gas is the difference between using it as a chemical substance for manufacture of other substances and burning it as a fuel for one-time use of its thermal energy. Since it is a finite resource that requires geologic time periods for its formation in large quantities, prudence might dictate that its long-term application be focused on its chemical value rather than its thermal energy value. Furthermore, since natural gas is a relatively friendly thermal energy resource for human habitats, its use for heating should be focused on residential use rather than the generation of electricity.

4.22 Growth of Natural Gas Consumption in the United States

A general picture of the growth in consumption of fossil fuels as an energy resource was developed in Chapter 3. A more detailed picture of growth in the consumption of natural gas specifically is needed to evaluate the intense competition for this fossil fuel in the future. Annual production of natural gas in the United States grew rapidly from about 100 BCF/a in the early 1900s to more than 12 TCF by 1960. The steep growth rate of marketed natural gas over this period (as given in Hubbert [3]) is shown in Figure 4-3.

The rate doubled to about 23.5 TCF/a by 2000 [4]. The disposition of natural gas since 1930 has shown an uneven growth history over the 70-year period. Figure 4-4 provides a composite picture of the variability of the mean annual growth rate over those decades. The input

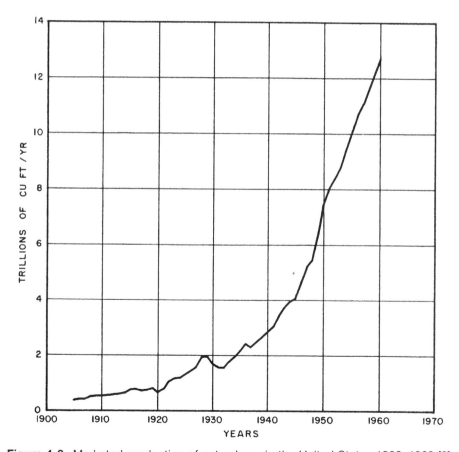

Figure 4-3. Marketed production of natural gas in the United States 1900–1960 [3].

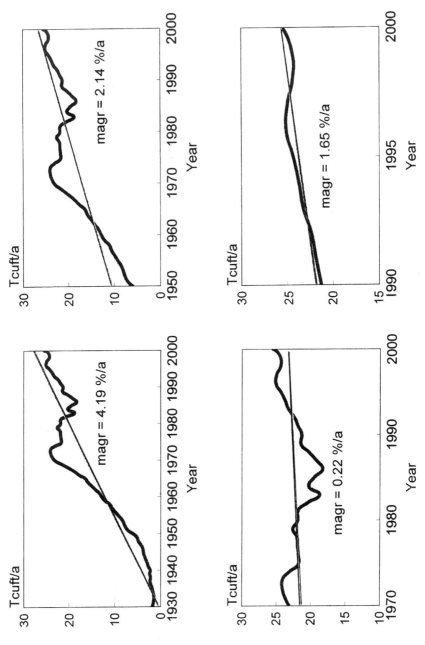

Figure 4-4. Mean annual growth rate of natural gas in the United States 1930–2000 [4].

data (including exports) from DOE/EIA [5] for the period are summarized in Table 4-2.

4.23 Forecast of Natural Gas Consumption through 2025

Table 4-3 shows the DOE/EIA forecast [4] for natural gas consumption in the United States in the period 2000–2025. The data are grouped into the four application sectors of residential, commercial combined with industrial, electricity generation, and transportation. The values modeled for transportation anticipate a steady growth at 10.4%/a only for compressed natural gas as an automobile fuel. The forecast does not anticipate a significant additional demand for natural gas as a feedstock for reformer production of hydrogen as a transportation fuel. This aspect of natural gas demand is examined in the discussion of primary resources for hydrogen fuel in Chapter 8.

4.24 Natural Gas Supply and Reserves

Through 2000, it generally was accepted that the supply (and price) of natural gas would remain the same well into the future. DOE/EIA [5] listed the supply of natural gas from 1985 to 2000, which was growing at an m.a.g.r. of 2.11%/a. DOE/EIA [6] listed the supply for 1988–1999, which grew at an m.a.g.r. of 1.85%/a. Table 4-3 shows a forecast m.a.g.r. value of 2.9%/a. Figure 4-5 shows the supply growth exceeding the demand growth of natural gas for both residential use and electric power generation. With demand growth for electric energy of 2.9%/a compared with demand growth for residential use of 1.1%/a for a total growth of 1.8%/a, the picture changes markedly through 2050, as shown in Figure 4-6, without the use of natural gas for hydrogen fuel production.

The extrapolation to 2050 indicates that at constant growth rates, the fraction of natural gas used for electric power generation, with an

Table 4-2 Consumption of natural gas in the United States 1930–2000 [5]

Period	Consumption (TCF/a)	m.a.g.r. (%/a)
1930–1950	1.87–6.02	6.67
1950–1970	6.02–23.1	6.20
1970–1990	23.1–21.3	−0.98
1990–2000	21.3–25.5	1.65

Table 4-3   Disposition of natural gas in the United States through 2025 [4]

Sector	Natural Gas Consumption (TCF/a) in Year						m.a.g.r. (%/a)
	2000	2005	2010	2015	2020	2025	
Residential	4.98	5.30	5.50	5.69	5.96	6.22	1.1
Commercial/industrial	11.5	11.7	12.5	13.4	14.3	15.3	1.5
Electricity generation	5.23	5.69	6.80	8.01	9.39	10.6	2.9
Transportation (CNG)	0.01	0.03	0.06	0.08	0.10	0.11	10.4
Total	23.5	24.8	27.1	29.5	32.1	34.9	1.8

m.a.g.r. twice that of the 1988–1999 value, could approach 40% of the natural gas supply. In contrast, the amount of natural gas for residential use through 2050 remains less than 15% of the natural gas supply. This contrast leads to the question raised in Table 4-1: How will your grandchildren afford to heat their homes?

Sustainability of natural gas supply depends on continuous provision of natural gas reserves. It was noted in Chapter 3 that the results from the early analysis by Hubbert [3] showed an ultimate resource of about 1000 TCF and current proved reserves of about 270 TCF. The data leading to these results are shown in Figure 4-7. For the approximate then-current production rate of 13.3 TCF/a, the R/P ratio was about 20 years.

Current estimates of natural gas reserves and R/P ratios, which are given in Table 3-7, are reserves of 184 TCF and an R/P ratio of 9.6 years. Undoubtably, additional discoveries of natural gas deposits will occur in future years as the demand rises, and the importation of natural gas (as liquid natural gas, LNG) undoubtedly will rise as the price of natural gas continues to rise. The use of LNG as a fuel in future motor

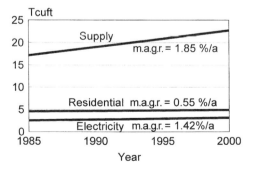

Figure 4-5. Growth of U.S. natural gas supply 1988–1999 [6].

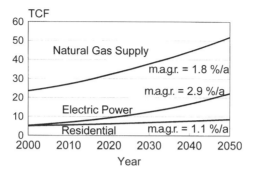

Figure 4-6. Growth of electric power and residential use of natural gas relative to supply.

vehicles may become important as economic and environmental concerns about petroleum-based fuels continue to increase.

4.3 NATURAL GAS COMMITMENT FOR ELECTRIC POWER GENERATION

Competition between sectors of the economy for natural gas may be considered to have started with the rapid installation of gas-fired tur-

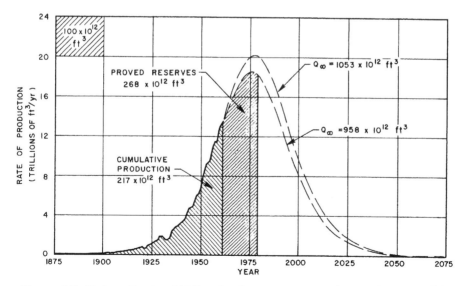

Figure 4-7. Early estimate of U.S. natural gas resource and proved reserves [3].

bines to meet the peaking demand for electricity. As was noted in Section 4.2, the fraction of natural gas consumption for electricity generation in 1985 was smaller than the fraction for residential use, but its rate of growth was three times greater over the decade. The DOE/ IEA 2002 energy forecast included an expectation that natural gas would be the preferred fuel for growth over the next 20 years at a growth rate of 5.2%/a, increasing from 15.6% to 32.7% of total generation. A major consideration for extended dependence on natural gas for long-term generation of electric power is the investment in gas-turbine installations, which require long-term amortization periods. The commitment may be expressed as

$$NGC = \sum N_{S,A} \, S_A \, CF \, (L_S - A) \qquad (4.2)$$

where NGC = natural gas commitment (GJ)
$\quad$ N = number of power plants of size S (MW) and age A (years)
$\quad L_S$ = expected mean lifetime for size S (years)
$\quad$ CF = conversion factor for plant factor (PF) [GJ/(PFxMWy)]

The commitment to natural gas is illustrated for the data (see Figure 4-8) on projected electricity generation capacity additions by DOE/ EIA [7]. The report indicated that natural gas units were expected to dominate new capacity additions compared with coal and renewables.

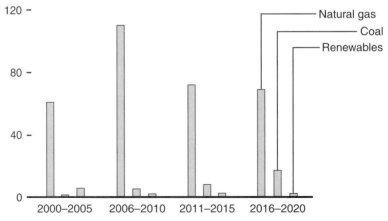

Figure 4-8. Projected additions to electricity-generating capacity (GW) 2000–2020 [7].

Table 4-4 shows the additions for natural gas units during the five-year periods, tabulated for natural gas consumption with a plant factor of 0.9 and an energy conversion factor of 3.345 CF/kWh.

The natural gas commitment through 2050 for only these capacity additions was calculated in spreadsheet form (Table 4-5) for a unit lifetime of L = 30 years during the five-year periods.

The total natural gas commitment for the DOE/EIA [7] projected electricity generation capacity additions for 2000 through 2020 with a 30-year lifetime per unit is shown in Figure 4-9. It is interesting to compare the 50-year natural gas commitment for the additional units projected to be installed from 2000 to 2020 (246 Tcuft) with the current proved reserves for the United States (184 Tcuft; see Table 3-7).

4.4 SUSTAINABILITY OF NATURAL GAS AS AN ENERGY RESOURCE

The forecast of natural gas demand for existing applications includes a small amount for the use of compressed natural gas (CNG), and liquid natural gas (LNG), as transportation fuels. It does not, however, include an allowance for large-scale demand for natural gas that would be needed as a feedstock if reforming of natural gas becomes a major source of energy for production of hydrogen as a fuel for fuel-cell vehicles. Although the use of hydrogen as a transportation fuel will be examined in the last three chapters, it is germane here to compare the characteristics of natural gas and hydrogen as transportation fuels. Table 4-6 compares the specific and volumetric energy of these two fuels in both compressed gas and liquid forms.

These data show that hydrogen yields almost 2.5 times more combustion energy per kilogram than natural gas. The data also show, however, that hydrogen, as the lightest gas in the periodic table, yields a

Table 4-4 Natural gas requirement for projected capacity additions

Period	GW by NG	PWh/a	Tcuft/a
2000–2005	60.4	0.48	1.6
2006–2010	109.3	0.86	2.9
2011–2015	71.9	0.57	1.9
2016–2020	69.1	0.54	1.8
Sum	310.7	2.45	8.2

Table 4-5   Natural gas commitment for projected capacity additions

Period	NG Commitment (Tcuft/5-year period)	Tcuft/a
2000–2005	1.6	1.6
2006–2010	1.6 + 2.9	4.5
2011–2015	1.6 + 2.9 + 1.9	6.4
2016–2020	1.6 + 2.9 + 1.9 + 1.8	8.2
2021–2025	1.6 + 2.9 + 1.9 + 1.8	8.2
2026–2030	1.6 + 2.9 + 1.9 + 1.8	8.2
2031–2035	2.9 + 1.9 + 1.8	6.6
2036–2040	1.9 + 1.8	3.7
2041–2045	1.8	1.8
2046–2050		0.0
Sum per 5-year period		49.2
Total commitment = 246 Tcuft		

very small energy content on a volumetric basis. This disadvantage is very important in considering gaseous hydrogen as a vehicle fuel. It means that a very large storage tank would be needed on-board to provide the driving range now obtained from 10 to 20 gallons of gasoline. Thus, hydrogen fuel would have to be supplied either as a liquid fuel or in very high-pressure tanks. But does it make sense to reform natural gas to hydrogen on the basis of their energy content?

Reforming of natural gas to hydrogen (which is described in Chapter 8) is accomplished by the overall reaction

$$CH_4 + 2H_2O \Rightarrow 4H_2 + CO_2 \tag{4.3}$$

which on a molar mass basis shows that 16 kg of methane produces 8 kg of hydrogen. Thus, the corresponding molar heat of combustion would be 222 kWh(th) (thermal energy) for methane and 266 kWh(th)

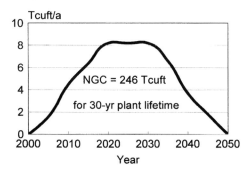

Figure 4-9. Natural gas commitment for projected U.S. additional units 2000–2020.

Table 4-6 Combustion energy of natural gas and hydrogen

Fuel Form	Specific Energy (kWh/kg)	Volumetric Energy (kWh/Nm³)
Natural gas		
Compressed (CNG)	13.9	3.38
Liquid (LNG)	13.9	5.80
Hydrogen		
Compressed (GH_2)	33.3	0.64
Liquid (LH_2)	33.3	2.36

for hydrogen as a fuel. These data raise the question, Why reform 16 kg of methane to produce 8 kg of hydrogen for a 20% increase in transportation energy rather than using it directly as a transportation fuel?

Much worldwide attention is being given to the potential for reforming natural gas to hydrogen for fuel-cell engine vehicles to replace internal-combustion engine (ICE) vehicles over the next 50 years. Methods for estimating the amount of natural gas that would be needed are discussed in subsequent chapters, but the amount is currently negligible and could grow to 19.5 TCF/a by 2050. With this value and the data from DOE/EIA [7] to 2020, Table 4-7 summarizes the long-term sustainability of natural gas in the United States by comparing demand in TCF/a to supply at constant m.a.g.r. to 2050.

The data in Table 4-7 indicate that on the basis of the assumptions made, the natural gas supply will be sufficient to meet demand through 2020 but that rapid growth in the manufacture of fuel-cell vehicles after

Table 4-7 Extrapolated natural gas demand and supply to 2050

Sector	Published Data 2000	Published Data 2020[a]	M.a.g.r. (%/a)	Extrapolated (2050)[b]
Commercial-industrial[c]	13.6	17.4	1.2	25.1
Residential	5.0	6.0	0.9	7.9
Transportation (CNG)	0.02	0.14	9.8	2.6
Electric power	4.2	10.3	4.5	40.0
Hydrogen production[d]	0.0	0.02	2.2	19.5
Total demand	22.8	33.8	3.4	95.1
Forecast supply[a]	22.7	34.1	2.1	64.4

[a] Adapted from DOE/EIA [7].
[b] Extrapolated at constant m.a.g.r.
[c] Includes use as petrochemical feedstock.
[d] All produced by steam reforming, none by electrolysis.

2050 will create an additional demand of 19.5 TCF/a, which is smaller than the calculated deficit of about 30 TCF/a between extrapolated demand and supply. Thus, some time between 2020 and 2050, the supply of natural gas will have to grow significantly or the demand will have to be reduced even without the availability of natural gas for production of hydrogen or its use directly as a substitute for petroleum fuels (gasoline, diesel, aviation fuel, etc.).

This brief discourse on the sustainability of natural gas as a desirable (green) energy source should provide sufficient reason for the government to look beyond 2025 in planning for the supply of efficient energy to continue the development of the U.S. economy. The only other large-scale sources of abundant energy besides fossil fuels are nuclear energy and renewable energy. These energy forms are discussed in Chapters 6 and 7 after an examination of the environmental impacts of energy utilization in Chapter 5. First, let us take a quick look at the potential for nonfossil fuels through the year 2025.

4.5 NONFOSSIL ENERGY RESOURCES

Curtailment of the use of fossil fuels for thermal energy because of resource depletion, the national economy, and environmental concern requires the substitution of alternative (nonfossil) energy resources. The two major sources of nonfossil fuel are nuclear energy (from thorium, uranium, and transuranic elements) and renewable energy (from incoming solar irradiance, primarily hydropower from the hydrologic cycle, wind energy from atmospheric thermal energy, and biomass energy from photosynthesis). These sources are described in Chapters 6 and 7. They already have become significant in electric energy generation.

4.51 Growth of Alternative (Nonfossil) Energy Use

Commercial generation of electricity by nuclear power began in the 1950s, and the number of nuclear power plants in the United States grew steadily through the 1980s, when public fears of nuclear power led to the cessation of further orders by the electric utility industry. Today some 104 nuclear power plants generate about 20 percent of the U.S. electricity supply. Although several countries have hesitated about building nuclear power plants, international development of commercial nuclear energy has continued to this date. By 2000, some 425 nuclear power plants in 31 countries were producing more than 3.0 PWh/a of electricity, representing 7.4% of the primary energy consumed and 20% of the electricity generated annually in the world.

Until the public desire for clean, green energy escalated in the 1960s, the major renewable energy resource was hydroelectricity, which reached near saturation on the large river networks in the United States and elsewhere. Geothermal energy (a renewable energy source only after hundreds to thousands of years of reheating the cooled rock deposits by thermal conduction in the earth's upper crust) was developed in the 1920s in Italy, New Zealand, the United States, Japan, and several other countries endowed with easily accessible deposits of thermal energy in rock formations and water or steam available to bring the heat to the surface in drilled wells. Today, renewable energy resources such as solar radiation, wind power, and biomass are in rapid commercial development. Table 4-8 shows the history of the use of nuclear and renewable energy resources for generation of electricity in the United States relative to total electricity generation. The increase in nuclear generation in the 1990s, with no new power plants, was due to increased plant efficiency. The percentage decrease in renewables has been due primarily to the saturation of hydropower relative to the growing total energy supply.

4.52 Forecast of Nonfossil Energy Supply

The annual energy outlook report prepared by the U.S. Department of Energy has slowly altered the forecast for nuclear energy over the last few years from the pessimistic view that commercial nuclear energy in the United States would be phased out by 2020 to a reluctant forecast that nuclear power capacity would be essentially constant through 2025. In the 2003 report [4], a modest growth of renewable resources (other than hydroelectric power) as a decreasing fraction of the total

Table 4-8 History of U.S. electricity generation by nonfossil fuels [8]

Year	Nuclear		Renewable		Total
	PWh	%	PWh	%	PWh
1980	0.251	11.0	0.285	12.4	2.290
1990	0.577	19.0	0.357	11.8	3.038
2000	0.754	19.8	0.357	9.4	3.802
m.a.g.r. (%/a)	5.51		1.82		2.82

electricity supply was forecast. Table 4-9 shows a summary of the 2003 outlook for nuclear and renewable energy relative to the historical values for 2001.

4.6 SUMMARY

The chapter examined the potential for extensive growth in energy supply to sustain economic development in the United States, especially under the constantly growing demand for electric energy and transportation fuel. The electronic age in the United States (and the rest of the world) is just beginning, and many new energy-intensive industries will be developed. The demand for energy will be accelerated by the recognition that civilization is vulnerable to terrorism and that much additional energy will be needed to provide a heightened degree of security for national populations and engineered structures. An analysis was made of the sustainability of fossil fuels, especially the more environmentally desirable natural gas, to meet these growing demands for additional energy. The problem will become more acute as hydrogen fuel is introduced to lessen reliance on increasingly expensive petroleum fuels over the next 50 years. Generation of electricity and the concomitant production of hydrogen fuel from natural gas do not appear to be feasible compared with their use for petrochemical synthesis and clean energy for residential use. The only other large-scale energy resources available to meet the growing demand are nuclear energy and renewable energy, which are explored in Chapters 6 and 7.

Table 4-9 Projections of nonfossil energy supply in the United States [4]

Year	Nuclear		Renewable		Total
	PWh	%	PWh	%	PWh
2001	0.769	22.8	0.358	10.6	3.370
2005	0.793	21.6	0.378	10.3	3.677
2010	0.800	19.5	0.393	9.6	4.105
2015	0.805	17.9	0.405	9.0	4.505
2020	0.807	16.5	0.416	8.5	4.887
2025	0.807	15.2	0.429	8.1	5.309
m.a.g.r. (%/a)	0.2		2.1		1.9

REFERENCES

[1] International Atomic Energy Agency and International Energy Agency, *Indicators for Sustainable Development,* 2001. Available at http://library. iea.org/dbtw-wpd/textbase/papers/2001/cds-9.pdf.

[2] C. Starr, "National Energy Planning for the Century: The Continental SuperGrid." *Nuclear News,* February 2002, pp. 31–35.

[3] M. King Hubbert, *Energy Resources: A Report to the Committee on Natural Resources.* National Academy of Sciences–National Research Council Publication 1000-D. Washington, DC, 1962.

[4] U.S. Department of Energy, Energy Information Agency, *Annual Energy Outlook.* Report No. DOE/EIA-0383(2003). Washington, DC: U.S. Department of Energy, 2004.

[5] U.S. Department of Energy, Energy Information Agency, *Natural Gas Annual 2000.* Report No. DOE/EIA-0130(00). Washington, DC: U.S. Department of Energy, 2001.

[6] U.S. Department of Energy, Energy Information Agency, *Annual Energy Outlook.* Report No. DOE/EIA-0383(01). Washington, DC: U.S. Department of Energy, 2000.

[7] U.S. Department of Energy, Energy Information Agency, *Annual Energy Outlook.* Report No. DOE/EIA-0383(02). Washington, DC: U.S. Department of Energy, 2001.

[8] U.S. Department of Energy, Energy Information Agency, *Annual Energy Review.* Report No. DOE/EIA-0384(02). Washington, DC: U.S. Department of Energy, 2003.

5

ENVIRONMENTAL
IMPACT OF
ENERGY CONSUMPTION

5.0 HISTORICAL PERSPECTIVE

Energy is defined as the ability to do work; the environment is defined as the surrounding material and cultural world. Since the 1960s, public concern about the *energy and environment crisis* has grown rapidly, based on the premise that overutilization of energy by an affluent society will result in severe degradation of the local and global environment. Much debate has taken place on whether energy and the environment are in conflict and whether there actually is a crisis. The global perspective became prominent as a result of international concern about changes in meteorological phenomena and their effect on global climate through the release of carbon dioxide from the combustion of fossil fuels. The local perspective has been an ongoing national and regional concern around the world with the large emissions of atmospheric pollutants into metropolitan air basins from electric power generation plants and automobile exhaust.

National awareness in the United States resulted in Congress passing the National Environmental Protection Act of 1969, which was signed into law January 1, 1970, as the first law of the new decade, in recognition of the importance of balancing the exponential growth of civilization with a consideration of its effect on the environment. It was claimed that for the first time in the history of industrial development, broad problems such as population growth, urbanization, resource exploitation, and appropriate technology were being taken into account for national planning purposes. Debate on whether energy use and the environment are in conflict and whether there is indeed an energy versus environment crisis has continued to this day.

With the mounting evidence around the world of acid rain effects from electric power generation, local air pollution from automobile emissions, and the potential for global climate change from the accelerating release of *greenhouse* gases into the global atmosphere, the world generally has accepted the philosophy that constraint of pollution emission is a desirable goal. A formal response was under way with the United Nations Framework Convention on Climate Change in 1992 and accelerated with the Kyoto Protocol adopted in Tokyo, Japan, in 1997. Science, technology, and social demand have led to vast research in the last 50-plus years on what environmental impact is, how to measure it, and how to set safety standards with and without economic restraints while continuing the development of human industry.

5.1 BASICS OF ENVIRONMENTAL IMPACT

The concept of environmental impact implies a harmful threat to the environment from some human activity carried out to gain an economic benefit. Human activities such as urbanization and transportation systems result in an environmental impact. Most of these activities do not result in harmful threat. As was noted in Axiom 2 of the human quest for abundant energy, people wish to live in comfort and ease in a safe and peaceful environment. The problem really becomes how to measure, in an "engineering quantitative" way, the absolute and/or relative importance of environmental impact of human activities when there is a choice of options and how to evaluate the consequences of each one in arriving at a decision about which option to choose. In the province of providing sufficient energy to a population, the options are what primary energy resources to use and how much of each is needed to produce the desired standard of living. Much environmental science has been developed in the past 50 years with respect to determining the environmental impact that results from each energy fuel cycle. The hope has been to arrive at an uncontested *optimum* energy plan such that there would be *sufficient* energy to satisfy some *preconceived* standard with *minimum* environmental impact. Unfortunately, there is much disagreement on the exact meaning of the words *optimum, sufficient, preconceived,* and *minimum* with respect to environmental impact, so that no clear way to select a unique energy plan has been available.

In the approach to an engineering method to measure and evaluate environmental impact, two parameters have become of major interest: magnitude and severity. Magnitude attempts to answer the question, How large is the impact? This is measured in terms of how much of a pollutant is released, how large the spacial extent (from local to global) is, and how long the time extent (from instant to forever) is. Severity attempts to answer the question, How serious is the impact? This generally is evaluated in terms of some relationship to a *maximum permissible standard* and involves the concepts of how many people are affected, the degree of physical damage, and the extent to which the harm is irreversible or recoverable.

5.11   Relationship between Magnitude and Severity

Evaluation of environmental impact requires a relationship between magnitude and severity, preferably in some continuous form of *cause and effect*. The relationship describes the chain of steps in the impact as follows:

$$\text{Release} \xrightarrow{\quad\text{Residual}\quad} \text{Impact} \xrightarrow{\quad\text{Exposure}\quad} \text{Threat}$$
$$\text{(action)} \qquad\qquad \text{(magnitude)} \qquad\qquad \text{(severity)}$$

Some examples of the components of the chain are as follows:

Action	Residual	Magnitude	Exposure	Severity
Coal burning	SO_2	Tons	Lungs	Choking
Nuclear power	^{131}I	dps	Thyroid	Cancer
Groundwater	Subsidence	Meters	Financial loss	Loss of home
Construction	Noise	dB	Ears	Loss of hearing

5.12   Consequences of Environmental Threat

The consequences of environmental threat are often difficult to define. They can range from a gradual lowering of environmental quality (sometimes difficult to observe) to ecological disaster (possible extinction of endangered species). The large variety of factors that affect the extent of severity include physical aspects (e.g., exposure rate and duration) and population aspects (e.g., individual age, occupation, and general state of health). Severity has two significant levels: the boundaries denoting a minimum level of exposure (the natural background) and a maximum level of exposure (the point where toxicity enters the range). A natural background exists in essentially all facets of the biosphere in which human activities add to the biological burden. These are pollution releases that occur in nature by geologic processes without any effort by humans. Some examples are as follows:

Natural Process	Pollutants Released
Volcanic eruptions	SO_2, H_2S, particulates
Forest fires	CO_2, ash particulates
Thunderstorms	O_3 (ozone), NO_x
Hydrologic cycle	Erosion, leaching of minerals

These releases result in a permanent minimum concentration of each pollutant in the environment, distributed geographically in space and time. The general concern about the severity of these releases center on the possibility of Darwinian changes over anthropological time from long-term exposure to the increasing natural background caused by small additions of these pollutants.

The toxicity boundary becomes the highest concern in the magnitude–severity relationship when the consequences reach the *acute* level of a readily observable impact on human health. At these levels, effects on individuals become population statistics. Impacts on natural cycles become of concern, as in the CO_2 cycle, in which the potential consequences for world climate change as a result of increasing CO_2 emissions from fossil fuel combustion to the global atmosphere have received worldwide attention. In the continuous search for quantitative means to measure and assess environmental impact, several categories of environmental burden and the resulting ecological consequences have become useful in establishing distinguishing levels of concern. A general view of these levels is shown in Figure 5-1.

5.13 A Hypothetical Example of Magnitude-Severity Analysis

One of the more important examples of environmental impact that was studied deeply during the early period of environmental science and engineering was the problem of acid rain that resulted from the release of sulfur dioxide (and nitrogen dioxide) from the smokestacks of coal-burning electric power plants that formed sulfuric acid (and nitric acid) in the atmosphere. Normal rain, containing its usual concentration of H_2CO_3 from atmospheric carbon dioxide, has a pH of about 5.5. On absorption of H_2SO_4 (and HNO_3), normal rain becomes acid rain, with its pH dropping to about 4.3. Deposition of acid rain (and wind-blown

Category of Ecological Consequence

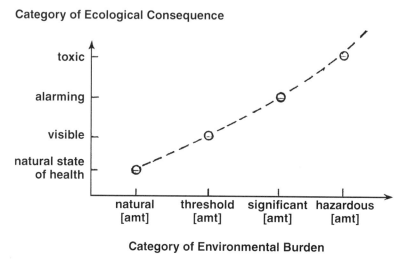

Category of Environmental Burden

Figure 5-1. General relationship between ecological consequence and environmental burden.

acidic particles) results in several unpleasant effects, including damage to forests and soils, aquatic and terrestrial animals, buildings and other structures, and human health.

The extensive literature on the magnitude-severity aspects of acid rain makes it very difficult to present a simple picture of the cause-effect relationship. Much less is known about the fate of the element fluorine, which is present in coal at a greatly reduced concentration compared with sulfur. As a very volatile element, it would be released more readily through smokestacks, and several levels of consequence of fluoride in drinking water are well known. Thus, it is an interesting hypothetical example for examining the magnitude-severity relationship of the environmental burden of fluoride emission from coal with respect to the threat of fluoride concentration in drinking water.

Figure 5-2 shows the linear slope resulting from continuous combustion of coal with a constant mean concentration of fluoride. The magnitude of fluoride released in kg/ton of coal is shown against the amount of coal burned in tons. The resulting concentration of fluoride in a drinking water supply would be a multiparameter function of the meteorological history during the path of the fluoride in the smoke emitted from the smokestack to the water supply. The figure shows a schematic of the potential magnitude (concentration of fluoride, $[F^-]$, in mg/L) resulting from variations in rainfall and the concentration in the water supply resulting from the mixing processes in the water body.

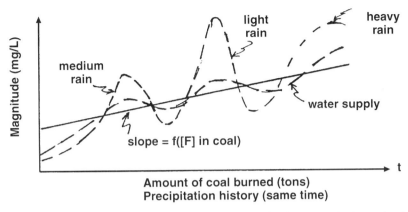

Figure 5-2. Fate of fluoride after emission from coal-fired power plant smokestacks.

The environmental burden of fluoride concentration in water supplies from electricity generation through the combustion of fluoride-containing coal is thus a function of several environmental physical processes (mostly meteorologic and hydrologic), including the following:

1. The volatilization of the fluoride content of the coal in the daily amount of coal used in the power plant and the amount released through the smokestack after any scrubbing processes
2. The washout and deposition of the fluoride after any chemical changes in the atmosphere and the occurrence and intensity of rainfall
3. The hydrologic mixing of the dry and rainfall deposition in the water supply body

The output of these processes yields a time-dependent history of the mean concentration of fluoride in a particular water supply. The severity of that concentration is determined by the long-term exposure in a population that uses that water supply. The severity–magnitude relationship is shown in Figure 5-3. The choice of fluoride in coal was made with an ulterior motive: the idea of perhaps illustrating what could be termed an environmental *benefit* from electric energy generation.

The natural concentration of fluoride in drinking water bodies has long determined the natural health severity of fluoride in water. The natural background is much less than 1 mg/L. Many communities in

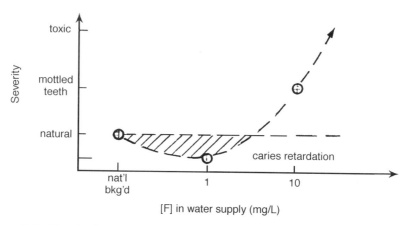

Figure 5-3. Magnitude-severity relationship of fluoride concentration in drinking water supply.

the United States and the world have chosen to add fluoride to the water supply to bring the concentration to 1 mg/L or more, up to the U.S. EPA maximum contaminant level (MCL) of 4 mg of fluoride per liter allowed in drinking water, because the dental profession "essentially" has shown that fluoride at that concentration decreases the incidence of caries in the teeth of children. However, it also is known that water with about 10 mg/L of fluoride can result in mottled teeth and that the water becomes toxic with a greatly increased fluoride concentration. Yet the practice of fluoridation is challenged continually. A newspaper headline in 1999 read "Mountain View (CA) is Sued over Fluoridation Plan: Lawsuit Meant to Block Treatment Project" [1]. Four years later, headline in Palo Alto (a neighboring town to Mountain View) read, "Petition Seeks to Take Fluoride out of Water" [2]. That year *Chemical & Engineering News* [3] reported, "Fluoride Concerns Surface Once Again." The uncertainty lingers.

5.2 THE SAGA OF THE GREENHOUSE EFFECT

Will the increasing concentration of CO_2 in the atmosphere change the global climate? During the last 50 years, abundant evidence has been collected that indicates a steadily increasing concentration of carbon dioxide in the global atmosphere that is attributed to the combustion of fossil (carbon-containing) fuels since the Industrial Revolution. It is accepted by most environmental scientists that the increase has been accompanied by a measured increase in mean atmospheric temperature that could affect the global climate markedly. Some scientists agree that the mean temperature of the earth's atmosphere is increasing but that this could be part of a natural long-term variation in atmospheric temperature as evidenced by the several ice ages that have occurred over geologic times that were not related to the emission of fossil fuel CO_2 into the atmosphere. Other scientists argue that the addition of CO_2 does indeed increase the temperature of the atmosphere above natural variations but that the effect on global climate will be negligible. Still other scientists think that even with changes in global climate, the severity to the environment will be minimal. Several books (e.g., [4],[5]) have explained "why we shouldn't worry about global warming."

How do we cope with this large uncertainty about the magnitude–severity chain that may be called the "Saga of the Greenhouse Effect"? One way is to examine the basic knowledge of the many components of carbon dioxide release to the atmosphere and the severity of global climate change processes. For an understanding of the total process, it is useful to examine each of the several parameters involved in the chain, which may be itemized as follows:

1. The natural atmospheric CO_2 cycle
2. The magnitude of the CO_2 concentration in the atmosphere
3. The magnitude of additional CO_2 release to the atmosphere
4. The magnitude of future CO_2 concentration in the atmosphere
5. Normal variations in the earth's climate
6. Greenhouse gas–imposed changes in atmospheric temperature
7. Severity of specific climate changes

5.21 Components of the Saga

1. The Natural Atmospheric CO_2 Cycle The global carbon cycle generally is portrayed as a "box diagram" in which the *stocks* of carbon in its major reservoirs on the earth are stored in boxes and transfers of carbon between those boxes are shown with vector arrows between them. Several summaries of current knowledge on the global carbon cycle have been published. An early one, prepared by the World Meteorological Organization (WMO) in 1977) [6], is shown in Figure 5-4.

Much of the literature on the global carbon cycle today is included in the reports of the Intergovernmental Panel on Climate Change (IPCC) [7] prepared in 1996 and the most recent multivolume publi-

The Atmospheric CO_2 Cycle*

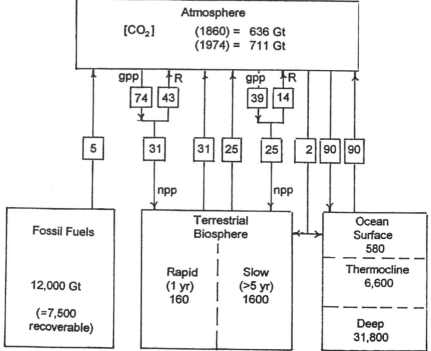

Exchangeable Carbon Reservoirs and Fluxes
Gt = 10^9 metric tons carbon in reservoir
□ = flux in 10^9 metric tons carbon per year
gpp = gross primary production; R = plant respiration; npp = net primary production
*From Report on Atmospheric Carbon Dioxide, WMO-474, 1977.

Figure 5-4. Global carbon-cycle box model [6].

cation in 2001 [8], with much deeper details about the science and data. The 2001 values for the box diagram are shown in Figure 5-5.
The IPCC divided the cycle into four parts:

1. The natural cycle
2. The human perturbation
3. Cycling in the ocean
4. Cycling on land

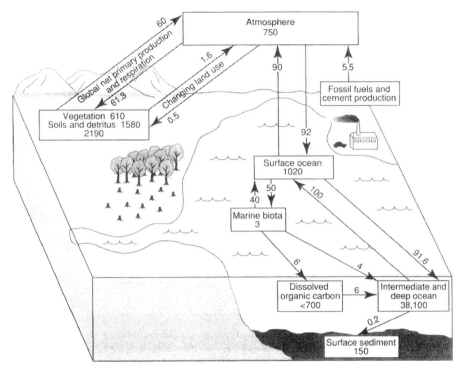

Figure 5-5. Global carbon-cycle box model [8].

The first two are shown in Figure 5-6. The large increase in knowledge acquired over the 25-year interval can be observed by comparing the two diagrams.

In particular, the estimated values of the atmospheric burden were noted in the 1977 WMO report as 636 Gt (gigatonnes of carbon; 1 Gt $= 10^9$ metric tons of carbon in the reservoir) in the year 1860, increasing by 75 Gt to reach 711 Gt in 1974. The corresponding value of carbon in the atmosphere in the IPCC report [8] from the preindustrial content increased to 730 PgC (petagrams of carbon; 1 Pg $= 10^{15}$ grams $= 10^9$ metric tons) through the 1990s. The IPCC [8] report noted that the increase in atmospheric CO_2 content was caused by anthropogenic emissions, with fossil fuel burning (plus a small contribution from cement production) releasing an average of 5.4 PgC per year in the 1980s and 6.3 PgC per year in the 1990s.

2. The Magnitude of the CO_2 Concentration in the Atmosphere
Physical measurements of CO_2 concentration $[CO_2]$ in the atmosphere (and inferences from indirect measurements) have shown that

94

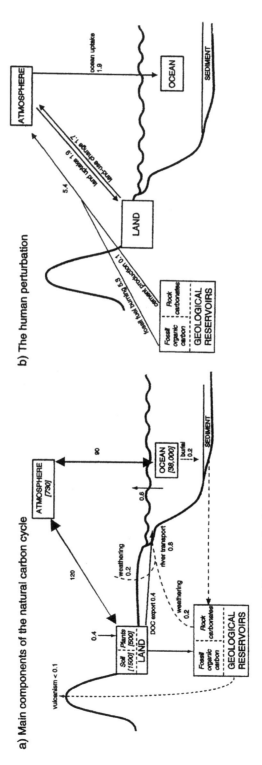

Figure 5-6. Global carbon cycle: (a) natural and (b) human perturbation [8].

the preindustrial era concentration has increased steadily over the era. The WMO'77 report [6] noted that mean [CO_2] was 295 ± 10 ppmv (parts per million by volume) in 1860 and 330 ppmv in 1974. Figure 5-4 shows that the atmospheric burden rose by (711 − 636) = 75 Gt in those 114 years (at a mean annual growth rate of 0.66 Gt/yr for a mean annual fossil fuel emission rate of 5 Gt/yr). The resulting change in CO_2 concentration was (330 − 295) = 35 ppmv in 114 years (at a mean annual growth rate of 0.3 ppmv/yr). Was the release of 75 GtC the cause of the atmospheric increase of 35 ppmv CO_2?

The IPCC'01 report [8] noted that the mean CO_2 concentration was 280 ± 10 ppm for several thousand years, as shown in Figure 5-7 for the period from A.D. 1000 to about 1750, and rapidly rose to a concentration of about 335 ppm in the twentieth century.

The most comprehensive series of physical measurements were made by Keeling and co-workers (e.g., [9]) continuously from 1958 in the isolated atmosphere atop the Mauna Loa volcano in Hawaii. The data (with seasonal variations) through 2004 (Figure 5-8) show the rise in concentration from about 315 ppmv in 1958 to about 380 ppmv in 2004 (at a mean annual growth rate of ~1.3 ppmv/yr). These data suggest that the atmospheric concentration of CO_2 will continue to rise in the future at an increasing growth rate if the flux of CO_2 to the atmosphere continues.

3. The Magnitude of Additional CO_2 Release to the Atmosphere

There is little doubt among most earth scientists that the atmospheric concentration of CO_2 has risen markedly since the Industrial Revolution. Questions linger about how much CO_2 has been released over that period. According to the WMO'77 report [6], the release rate

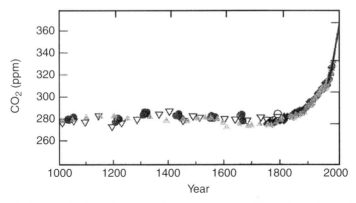

Figure 5-7. Atmospheric CO_2 concentration from A.D. 1000 through A.D. 2000 [8].

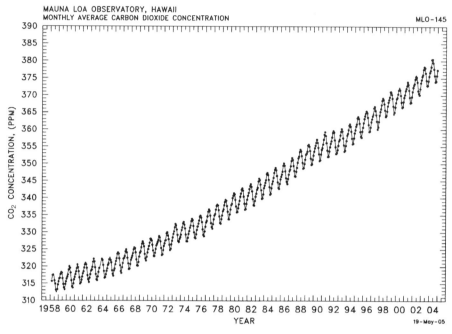

Figure 5-8. Atmospheric CO_2 concentration at Mauna Loa Observatory 1958–2004 [9].

of CO_2 (see Figure 5-4) was 5 Gt/yr (PgC/yr), with an integrated total release from the fossil fuel compartment of 136 Gt. The increase of CO_2 in the atmosphere was given as 75 Gt (PgC), which indicates that only about 55% of the CO_2 release remained in the atmosphere. Thus, the question could be raised: Where is the rest? Many of the studies in the IPCC reports dealt with this question. The IPCC'01 report [8] showed (see Figure 5-6b) that the current release rate from fossil fuel burning (5.3 PgC/yr) and cement production (0.1 PgC/yr) is 5.4 PgC /yr, which is in agreement with the WMO'77 estimate. The net uptake rate by land is the "residual terrestrial sink" [the net primary production (NPP) of biomass minus carbon losses caused by respiration and fire] of 1.9 PgC/yr offset by the land release processes (e.g., deforestation and land use changes) of −1.7 PgC/yr, yielding a very uncertain difference of 0.2 ± 0.7 PgC/yr. The net uptake by the oceans, which is governed primarily by ocean circulation and carbonate chemistry, is shown as 1.9 PgC/yr. The recent values of 5.4 PgC/yr released minus the returns of about 2.1 PgC/yr to the land and ocean compartments yield a net release of 3.3 PgC/yr, accounting for 61% of the release, which also is in agreement with the 55% estimate in 1977.

4. The Magnitude of Future CO_2 Concentration in the Atmosphere
A more difficult problem in evaluating the magnitude-severity relationship of greenhouse gas emission, which has been examined mostly for CO_2, is estimating the magnitude of emissions far enough into the future that the severity becomes more than just readily visible. Here again, the two reports show considerable advances. The WMO'77 report extrapolated the possible future CO_2 atmospheric burden in three scenarios to 2025: (1) business as usual, (2) most likely, and (3) with zero population growth. The relative changes in slope relative to the historical growth rate of 0.3 ppmv/a from 1860 to 1974 for the three models are shown in Figure 5-9. The range in 2025 would be from 425 to 530 ppmv at m.a.g.r. from 0.5 to 1.0%/a. The m.a.g.r for the most likely concentration of 460 ppmv would be 0.65%/a, about twice that of the early [6] period.

The corresponding forecasts in the IPCC'01 report [8] were made with two sophisticated models that evaluated six emission scenarios to project future CO_2 concentrations through the twenty-first century. Those scenarios covered an integrated framework of selected socioeconomic development paths for which the emissions and concentrations of greenhouse gases and aerosols were forecast; those results then were used to estimate the potential climate changes, such as temperature rise and sea-level rise, and finally were used to forecast the impact on

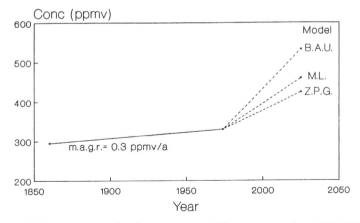

Figure 5-9. Forecast growth of atmospheric CO_2 concentration 1974–2025 [6].

human and natural systems. The range of increase in CO_2 concentration projected by the two models for the six scenarios was from 540 to 970 ppm, as shown in Figure 5-10. The wide difference in growth rates reflects the wide differences in assumptions on the emphasis placed on economic versus environmental goals and the physical extent of regional versus global coverage. The work by the Intergovernmental Panel on Climate Change represents a summary of the worldwide re search effort focused on the problem of the magnitude-severity analysis of greenhouse gas emission from fossil fuel combustion and other sources. The study is well worth the interest of readers of this book; the major publications of the IPCC program can be read at //www. ipcc.ch.

5. Normal Variations in the Earth's Climate   The magnitude of the problem of greenhouse gases may be accepted as reasonably well known. The increase in CO_2 concentration in the atmosphere since the Industrial Revolution has been well measured. The second half of the problem—the magnitude of the resulting climate change—also must be examined in detail. This half of the problem is more difficult in that the earth's climate may be described by some as an artist's conception and by others as a statistical Pandora's box. This is the case because although climate is a part of meteorological sciences, the database is very limited in terms of physical measurements in time and over ge-

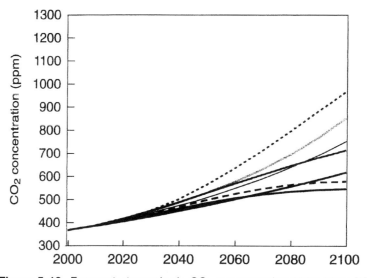

Figure 5-10. Forecast atmospheric CO_2 concentration 2000–2100 [8].

ography. Thus, the range of normal variation in the earth's climate is uncertain. Some of the parameters affecting the earth's climate include

The astronomical setting: variations in solar radiance
Chemical composition and particle loading
    moisture content: driving force of the hydrologic cycle
    CO_2 content: radiative balance of the atmosphere
    Particulates: condensation nuclei and reflectivity
Hydrodynamics and thermodynamic interactions with oceans, ice masses, land surfaces, and human perturbations
Each of these factors is function of time and natural time cycles.

Thus, the problem becomes how to characterize observed changes in this list as being due to either natural phenomena with constant (or variable) time periods or human activity. The measurement of climate change with respect to any climate-affecting parameter must be interpreted with respect to the equation of their relationship evaluated on the basis of their uncertainties:

$$\Delta \text{Climate} \pm \sigma(\text{change}) = f(\Delta \text{Parameter} \pm \sigma(\text{parameter}))   (5.1)$$

Thus, for each component of climate (atmospheric temperature, precipitation patterns, ocean temperature and mixing rate, melting rate of ice masses, raising of the sea level), four values must be obtained and compared: the magnitude of the climate component change, the uncertainty of the magnitude, the magnitude of the change in the forcing parameter, and the uncertainty of its magnitude. This constitutes a major component of the science of climate.

6. Greenhouse Gas–Imposed Changes in Atmospheric Temperature   Sections (3) and (4) in the Saga have shown that the atmospheric CO_2 concentration has risen over two centuries and will continue to rise in the future. IPCC'01 [8] notes that the global average surface temperature has increased by $0.6 \pm 0.2°C$ since the late nineteenth century. Figure 5-11 shows the combined annual land-surface air and sea surface temperature anomalies from 1861 to 2000 relative to the period 1961–1990. The bars on the annual numbers are two standard error uncertainties. Forecasts of temperature rise during the twenty-first century based on several models [8] are shown in Figure 5-12.

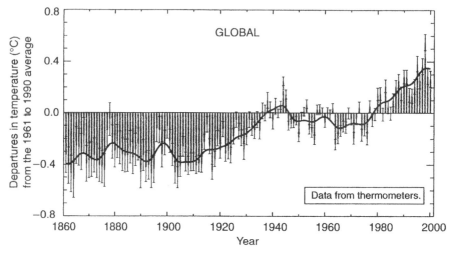

Figure 5-11. Global surface temperatures 1861–2000 relative to 1961–1990 [8].

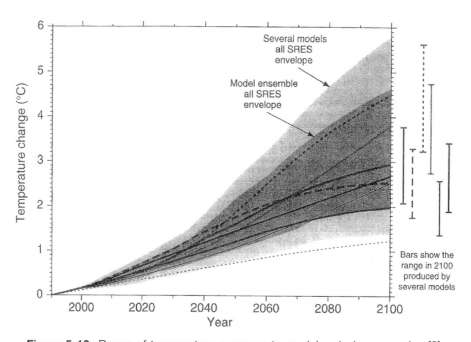

Figure 5-12. Range of temperature response to model emission scenarios [8].

The expected increases in temperature range from more than 1°C to almost 6°C. The resulting severity of any of the possibilities for climate change would be a function of the actual magnitude of further emissions of greenhouse gases under future worldwide regulatory agreements.

The report also notes that balloon and satellite records since 1979 show that the troposphere has warmed and that the stratosphere has cooled. These changes of temperature below and above the tropopause are in accord with the expectation of the effect of the emission of greenhouse gases and particulates in the atmosphere. The relative importance by Eq. 5.1 of the two components affecting the atmosphere's solar irradiance are given by the magnitude of the two world temperature gradients: $\Delta T_w / \Delta [CO_2]$, which is positive in the troposphere by the greenhouse effect and $\Delta T_w / \Delta [particulates]$ which is negative by the increased reflectivity of the incoming solar radiation.

7. The Severity of Specific Climate Changes   It is important to correlate the change in world climate with changes in the emission of greenhouse gases and particulate matter. However, the severity of changes in world climate, independent of the causes, is also of great importance in considering the effect on the world environment. The Saga of the Greenhouse Effect will continue with current and future research, as summarized in the IPCC program, that should reveal increasing knowledge about the effects of climate changes in the form of the following questions:

Thermal effects: Will warming of the atmosphere, oceans, and continents continue?

Biological effects: Will there be a change in photosynthesis?

Marine effects: Will the oceans be the long-term sink for $\Delta [CO_2]$?

5.3   LOCAL AIR POLLUTION FROM AUTOMOBILE EXHAUST

Although the magnitude-severity analysis of global atmospheric pollution is difficult to assess, the problem of metropolitan air pollution has been well documented worldwide. The distribution of air pollution problems noted by the World Health Organization [10] in the major cities of the world for the major air pollutants sulfur dioxide (SO_2), suspended particulate matter (SPM), lead (Pb), carbon monoxide (CO), nitrogen dioxide (NO_2), and ozone (O_3) is shown in Figure 5-13. The

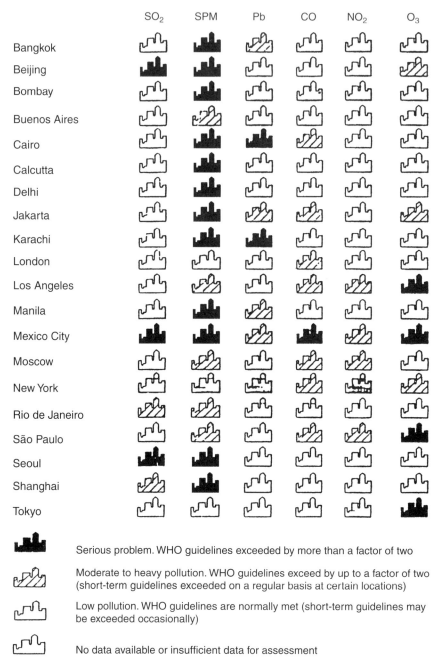

Figure 5-13. Distribution of worldwide air pollution problems [10].

solid figures are for cities that exceeded WHO guidelines by more than a factor of two.

The specific problem of ozone concentration (the precursor to smog) in the five most smog-polluted mega-cities in the world reported by the United Nations Environment Programme [11] is illustrated in Figure 5-14. The data for years 1989 to 1991 show that the ozone concentration in Mexico City was *not* above the standard for about 15 days per year.

Historical data for air pollution in the United States are compiled by the U.S. Environmental Protection Agency (EPA). The trend in air pollution for the last 30 years may be seen in its publications of 1972 and 2002. Air pollution emissions of the five largest components from 1940 to 1970 [12] are listed in Table 5-1.

The pollutants are carbon monoxide (CO), sulfur oxides (SO_x), suspended particulate matter (SPM), volatile organic compounds (VOC), and nitrogen oxides (NO_x). The change in the magnitude of these emissions from 1970 to 1998 in relation to efforts to reduce air pollution in several sectors of the economy are noted in Table 5-2. The distribution by source is listed in million tons/year.

5.31   Environmental Impact of Smog

The historical debate of energy versus environment has focused on two major sectors of the economy: stationary electric power plants and vehicular transportation. The concern about power plant emissions has changed over the years from the regional problem of acid rain deposition to the global problem of climate change that might result from the continued release of large quantities of carbon dioxide (and other greenhouse gases) from the combustion of fossil fuels. The emission of tailpipe gases from motor vehicles also has been a long-standing concern, but on a smaller scale, focused on the release of the precursor gases that form "smog" in metropolitan air basins in which extensive urban development has resulted in *traffic jams* and serious air pollution. The two more important contaminants from the combustion of fossil fuels in motor vehicles that form smog are nitrogen oxides (NO_x) and volatile organic compounds (VOC). Since volatile organic compounds also come in large quantities from nonvehicular activities and the formation of nitrogen oxides results mainly from high-temperature combustion of fuel, it will be instructive to focus on the nature of NO_x in urban air pollution.

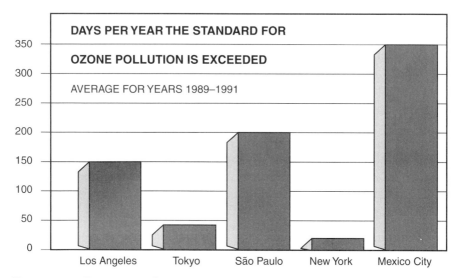

Figure 5-14. Frequency of ozone pollution above standard for years 1989–1991 [11].

5.32   Nitrogen Oxides in Photochemical "Smog"

In the natural state of "normal" (down-to-earth) dry air, the concentration of nitrogen oxide (NO) ranges from 0.25 to 0.5 ppmv and the concentration of nitrogen dioxide (NO_2) ranges from 0.001 to 0.002 ppmv. Formation of nitrogen oxide generally occurs in air by nitrogen "fixation," which is the reaction of air nitrogen and oxygen by

$$N_2 + O_2 \Rightarrow 2NO \tag{5.2}$$

and by high-temperature combustion

Table 5-1   Air pollution emissions in the United States 1940–1970 [12]

Pollutant	Mass (million tons)
CO	85–150
SO_x	22–34
SPM	25–27
VOC	19–35
NO_x	7–23

Table 5-2   Distribution of air pollutants by source for 1970 compared with 1998 [12, 13]

	CO		SO_x		SPM		VOC		NO_x	
Source	1970	1998	1970	1998	1970	1998	1970	1998	1970	1998
Transportation	111.0	70.2	1.0	0.4	0.7	0.7	19.5	7.8	11.7	13.0
Electric power	0.8	5.4	26.5	16.7	6.8	1.1	0.6	0.9	10.0	10.2
Industry	11.4	3.6	6.0	1.5	13.1	0.6	5.5	1.4	0.2	0.8
Solid waste	7.2	1.2	0.1	0.1	1.4	0.3	2.0	0.4	0.4	0.1
Other	16.8	9.1	0.3	0.9	3.4	32.0	7.1	7.4	0.4	0.3
Total	147.2	89.5	33.9	19.6	25.4	34.7	34.7	17.9	22.7	24.4

$$N + O \Rightarrow NO \tag{5.3}$$

The formation of nitrogen dioxide results from the oxidation of NO by both a slow reaction with oxygen:

$$2NO + O_2 \Rightarrow 2NO_2 \tag{5.4}$$

and a fast reaction with ozone:

$$NO + O_3 \Rightarrow NO_2 + O_2 \tag{5.5}$$

The formation of smog occurs by photochemical reactions with nitrogen dioxide, producing a chain reaction with atomic oxygen and VOC hydrocarbons that results in the formation of chemically reactive free radicals ($R^\bullet$ and $OH^\bullet$) by the transfer of a hydrogen atom from the VOC to the oxygen atom:

$$
\begin{aligned}
NO_2 + h\nu &\Rightarrow NO + O \\
O + RH &\Rightarrow R^\bullet + OH^\bullet \\
OH^\bullet + RH &\Rightarrow R^\bullet + H_2O
\end{aligned}
\tag{5.6}
$$

where $h\nu$ = photons of ultraviolet light (from sunshine)
RH = hydrocarbons (from VOCs)

Typical concentrations (in ppmv) of the gases in photochemical smog are listed in Table 5-3.

Table 5-3 Typical concentration of gases in
photochemical smog

Component	Concentration (ppmv)
Major gases	
H_2O	2×10^6
CO_2	4×10^4
CO	4×10^3
CH_4	250
Smog gases	
NO_x	20
O_3	50
VOC	10–60

5.33 Magnitude–Severity Aspects of Nitrogen Oxides

Magnitude (NO_x Emissions from Motor Vehicles)   The magnitude of NO_x emissions in the United States was listed in Table 5-2. In 1970, the fractional emission was 51.5% from the transportation sector and 44.0% from electric power plants [11]. In 1998, NO_x emission was 53.3% and 41.8%, respectively, from these sources, indicating that as total energy consumption increases, the smog problem will not lessen greatly if transportation sector use of fossil fuels continues.

Severity (Health Effects from "Hazardous Air Pollutants")   U.S. EPA [13] noted that health effects which may be caused by hazardous air pollutants (HAPs) include cancer, neurological, cardiovascular, and respiratory effects, effects on the liver, kidney, immune system, and reproductive system, and effects on fetal and child development. More than half of the 188 HAPs have been classified by U.S. EPA as 'known,' 'probable,' or 'possible' human carcinogens.

5.4 VALUE OF AIR QUALITY IMPROVEMENT IN TRANSPORTATION

The health effects of air pollution have been studied by the medical profession in great detail. For purposes of providing an engineering view of this key problem in the quest for abundant energy, it is useful to examine the severity of the problem in numeric terms, if possible, by investigating the value of air quality improvement. A model that attempts to do this would have to consider the major factors listed below (with appropriate symbolism):

Factor	Notation
Vehicle emissions	E_i (e.g., Mt/yr)
Pathway dispersion	D_j (by process)
Population exposure	P_k (e.g., g/cap)
Health effects	H_l (number/yr)
Health value	V_1 (e.g., ¢/mile)

Vehicle Emissions   The annual emission rate of pollutants is

$$E_i = 10^6 \sum N_i \; EF_{i,j} \; M_j \tag{5.7}$$

where E_i = emission rate of pollutant i (Mt/yr)
  i = NO_x, SO_x, PM_{10}, . . .
  N_j = number of vehicles of type j (in millions)
  j = autos; SUVs, vans, buses, trucks, . . .
  $EF_{i,j}$ = emission factor of pollutant i from vehicle type j (g/mi)
  M_j = mean annual distance travel in vehicle type j (mi/yr)

Also, $N_j \; M_j$ = VMT (vehicle miles travel per year) for M_j in miles
  $N_j \; M_j$ = VKT (vehicle kilometers travel per year for M_j in km.

Pathway Dispersion   The function for the dispersed concentration $[C]_j$ of pollutants is

$$[C]_j = f(D_j \; (p, x, t) \tag{5.8}$$

where D_j = dispersion due to process p at location x at time t
  with j = diffusion   (stationary air)
      convective transport (air flow and mixing)
      chemical reactions (e.g., smog formation)
      removal processes (rainout washout, etc.)

Population Exposure   The exposure-dose relationship for pollutants is

$$D_i \; (x,t) = \sum P_k \; [C_i] \; (x,t) \; F(i, k, X, T) \tag{5.9}$$

where D_i (g/cap) = dose received by population in air basin for each population group k exposed to air concentration of pollutant i with dosimetry factor F for exposure pathway X during exposure time T per day

$\quad\quad\quad\quad$ k = age, gender, . . . , groups

$\quad\quad\quad\quad$ X = inhalation, ingestion (water, food), adsorption, . . .

$\quad\quad\quad\quad$ T = f(fraction time indoors, outdoors, . . .)

Health Effects   The number of health effects per year is

$$H_{i,j} = \sum D_{i,k} \, R_{k,l} \quad\quad\quad (5.10)$$

where $H_{i,j}$ = number of health effects of type l as a function of dose received by population of organ k with probability R incurring health effect l

$\quad\quad\quad$ l = lung cancer, asthma, eye irritation, . . .

Health Value   The economic value of reducing vehicle emissions, for example, by replacing fossil fuels (ff) with hydrogen fuel (H_2), is

$$V_i = \sum (H_{i,j} \, (ff) - H_{i,j} \, (H_2)) \, W_i \quad\quad\quad (5.11)$$

where W_i = weighting factor of relative severity of ailment l expressed in cost units such as \$/cap or ¢/mile.

5.5   SOME DATA FOR THE LOS ANGELES AIR BASIN

One of the more air-polluted (and studied) cities in the United States is the metropolis of Los Angeles, which is part of the South Coast Air Quality Management District in California. Health impacts have been recorded and evaluated for the air pollutants and their health impacts:

Pollutant	Health Impacts
[NO_x]	Respiratory problems and lung damage
[SO_x]	Respiratory problems and lung damage
[CO]	Toxic through lack of oxygen
[Ozone]	Respiratory problems and lung damage, asthma, eye irritation, nasal impairment
[Particulates]	Bronchitis, eye, nose, and throat irritation
[Lead]	Brain damage

An early estimate of the economic value of reducing air pollution in the metropolitan Los Angeles air basin was published in 1995 (Lloyd [14]) based on the medical data of Hall and associate in 1992 [15], with the following results:

Estimated value of reducing air pollution	$9.8 billion per year [15]
Annual vehicle travel data	VMT $= 1.1 \times 10^{11}$ mi/yr [14]
Based on 50% air pollution from motor vehicles,	
Epidemiological value	4.5 ¢/mi.

A more current evaluation of the value of reducing air pollution through the replacement of fossil fuels with hydrogen fuel is reviewed in Chapter 10.

5.6 SUMMARY

The chapter introduced the basic concept of environmental impact, with an engineering method to distinguish between the parameters of magnitude (how much) and severity (how bad). The relationship between them was examined in terms of environmental burden and ecological consequence. An example was the public's uncertainty about fluoridation of drinking water and toothpaste, which continues today. The magnitude and severity aspects of greenhouse gas emission and world climate change on a global scale and automobile exhaust of nitrous oxides and the epidemiological health effects of smog on an urban scale were examined. The externality value of air pollution abatement by hydrogen fuel is examined in Chapter 10.

REFERENCES

[1] *San Francisco Chronicle,* December 15, 1999.

[2] *Palo Alto Daily News,* May 17, 2003.

[3] B. Hileman, "Fluoride Concerns Surface Again," *Chemical & Engineering News,* August 7, 2003.

[4] R. L. Bradley, Jr., *The Increasing Sustainability of Conventional Energy.* Policy Analysis No. 341. Washington, DC: Cato Institute, April 1999.

[5] T. G. Moore, *Climate of Fear: Why We Shouldn't Worry about Global Warming.* Washington, DC: Cato Institute, 1998.

[6] World Meteorological Organization, *Report on Atmospheric Carbon Dioxide.* WMO-474, 1977.

[7] J. T. Houghton et al., *Climate Change: 1995: The Science of Climate Change.* IPCC Report. Cambridge, UK: Cambridge University Press, 1996.

[8] Intergovernmental Panel on Climate Change, *Climate Change 2001: The Scientific Basis.* Cambridge, UK: Cambridge University Press, 2001.

[9] C. D. Keeling and T. P. Whorf, *Atmospheric Carbon Dioxide Record from Mauna Loa 1958–2004.* MLO-145, May 19, 2005. Available at cdiac.esd.ornl.gov/trends/co2/sio-mlo.htm.

[10] World Health Organization, Geneva, Switzerland.

[11] United Nations Environment Programme. *Urban Air Pollution in Mega Cities of the World.* Nairobi, Kenya: UNEP, 1992.

[12] U.S. Environmental Protection Agency, *National Air Pollutant Trends.* EPA 454/R-00-72, Washington, D.C. 1972.

[13] U.S. Environmental Protection Agency, *National Air Pollutant Trends.* EPA 454/R-00-002, Washington, D.C. 2002.

[14] A. C. Lloyd, "The Role of Hydrogen in Meeting Southern California's Air-Quality Goals." Proceedings, International Hydrogen and Clean Energy Symposium '95, Tokyo, Japan, February 1995.

[15] J. V. Hall, A. M. Winer, M. T. Kleinman, F. W. Lurman, V. Brajer, and S. D. Colome, "Valuing the Health Benefits of Clean Air." *Science* 225: 812–817, 1992.

6

THE NUCLEAR
ENERGY ERA

6.0 HISTORICAL PERSPECTIVE

The nuclear age began about 4 billion years ago with the creation of the earth. Among the original 90-plus chemical elements that constitute the earth's crust, many (hydrogen, carbon, potassium, uranium, and others) are radioactive, meaning that they can disintegrate spontaneously into other chemical elements. The sun, the source of essentially all of the earth's energy, is a stellar thermonuclear reactor that radiates nuclear energy throughout the solar system from the nuclear reactions that take place continuously in its sphere. Natural nuclear reactors existed on earth about 2 billion years ago and lasted for about 500,000 years. How this could have occurred and how we know it occurred are described in this chapter. Nuclear energy as it is known today began in the late nineteenth century, was highlighted in 1905 with the Einstein equation $E = mc^2$ that related mass and energy, and steadily developed in the first half of the twentieth century with the development of atomic and hydrogen bombs as military weapons and the initiation of the nuclear energy age with the construction of nuclear-fission power reactors and the search for controlled thermonuclear power reactors that in the future could duplicate the sun's energy on earth.

Electricity from nuclear power reactors could play an important role in worldwide production of hydrogen as a sustainable and environmentally clean fuel. The technology for construction and operation of large-scale safe nuclear power reactors is well established, and nuclear reactors are now in operation in most of the developed counties of the world, producing from a few percent to more than 70 percent of their national electricity demand.

This chapter reviews the technology for using nuclear energy to assist in the large-scale production of hydrogen by electrolysis and possibly, in the future, by thermal decomposition of water. The review includes an "engineering" model of the atom, with its nucleus consisting solely of protons and neutrons, surrounded by orbiting electrons. The model describes the science of nuclear energy sufficiently to explain the technology for generating abundant electric energy by nuclear fission. The model covers the basic structure of the atomic nucleus, isotopic composition and abundance, atomic mass, and the equivalence of mass with energy by the Einstein equation from the calculation of binding energy, which is the source of nuclear energy, both from nuclear fission and from thermonuclear fusion. The model covers aspects of nuclear stability and the important types of radioactivity and radiation.

The chapter also reviews the technology of nuclear power plants (both existing and future designs) and the potential for nuclear power plants to provide both electric power and hydrogen fuel. Operating such plants over the daily changing demand for electric power at constant conditions (baseload operation) results in greater efficiency and would provide both the electric energy and the hydrogen fuel at least cost. The book concludes in Chapter 10 that isolated energy parks could synergistically combine nuclear and renewable energy resources for maximum safety and efficiency and the lowest cost of electric energy and hydrogen fuel.

6.1 BASIC ELEMENTS OF NUCLEAR SCIENCE

It has been convenient in describing the science of the atomic nucleus to engineering students to use an "engineering" approach in which nuclear structure is modeled as a tightly bound nucleus containing solid, hard spheres of protons and neutrons surrounded by orbiting electrons. This model neglects the many decades of progress in meson, quark, and strange-particle physics research that are needed to understand nuclear physics, but the model does allow a description of the nuclear physics needed to explain nuclear phenomena that relate to the ability to construct nuclear power plants for generation of electricity on a massive basis.

6.11   The Atomic Nucleus

Under the *engineering* model of the atom (Figure 6-1), the structure of an atomic nucleus is described as a tightly bound nucleus containing two types of nucleons:

1. Protons (p^+): The atomic number (Z) of the atom is the number of protons in the nucleus, which determines the chemical identity of the atom.
2. Neutrons (n): The mass number (A) of the atom is determined by the number of protons (Z) and the number of neutrons (N), with A = Z + N.

The nucleus is surrounded by

3. Electrons (e^-): The neutral atom has Z orbital electrons; the outermost (valence) electrons determine the chemical behavior of the atom.

6.12   Isotopic Composition and Abundance

The identity of each element in the periodic table is determined by its atomic number (Z). Most of the elements exist in nature with two or

Figure 6-1. Engineering model of an atomic nucleus with its surrounding electrons.

more allowable combinations of nucleons as isotopes of each element. The isotopic composition (Z, N) is determined by the mass number of the isotope (A = Z + N). Other isotopes of the elements can be produced artificially by high-energy machines; most of these isotopes are radioactive. The isotopic abundance (f) of the isotopes in each element in the periodic table is expressed as the fraction found in the natural element. Twenty of the elements [e.g., sodium (Z = 11, N = 12)] exist with only one isotope (^{23}Na); for these elements, the isotopic abundance of the one natural isotope is f = 1.000.

Hydrogen, atomic number 1 in the periodic table, has one proton in the nucleus, but the element hydrogen (Z = 1) exists in nature with three isotopic compositions:

N = 0 ^{1}H Normal hydrogen Stable   = Hydrogen   f = 0.99+
N = 1 ^{2}H Heavy hydrogen _ Stable   = Deuterium f = <0.01
N = 2 ^{3}H Heaviest hydrogen Radioactive = Tritium

Carbon, atomic number 6, has six protons in the nucleus, but the element carbon (Z = 6) exists in nature with four isotopic compositions:

N = 5 ^{11}C Radioactive
N = 6 ^{12}C Stable   f = 0.99+
N = 7 ^{13}C Stable   f = <0.01
N = 8 ^{14}C Radioactive

Uranium, atomic number 92, has 92 protons in the nucleus, but uranium (Z = 92) exists in nature with three isotopic compositions:

N = 142 ^{234}U Radioactive f = 0.000061
N = 143 ^{235}U Radioactive f = 0.0072
N = 146 ^{238}U Radioactive f = 0.9927

6.13 Atomic Mass

The standard for atomic mass of the elements was adopted internationally in 1961, when an atom of ^{12}C was defined as having a mass of exactly 12 atomic mass units (amu). The mass in grams is calculated with the use of Avogadro's number, $N^\circ = 6.022 \times 10^{23}$ atoms/g-amu:

$$1 \text{ amu} = \frac{1}{6.022 \times 10^{23}} = 1.66 \times 10^{-24} \text{ g} \qquad (6.1)$$

Thus, a single ^{12}C atom has a mass of 2×10^{-23} g.

The masses of the basic atomic particles are as follows:

Proton	p^+	$= 1.0072765$ amu
Neutron	^1n	$= 1.0086649$ amu
Electron	e^-	$= 0.0005486$ amu
Hydrogen atom	^1H	$= 1.0078252$ amu

6.14   Equivalence of Mass and Energy

In 1905 Einstein, in developing his special theory of relativity, came to the conclusion that the properties of mass (m) and energy (E) are equivalent and can be expressed by the equation

$$E = mc^2 \qquad (6.2)$$

where c is the speed of light (3.0×10^{10} cm/sec in vacuum). It follows from Eq. 6.2 that potential energy is stored as mass and that in a given system a change in mass is accompanied by an equivalent change in energy:

$$\Delta E = \Delta mc^2 \qquad (6.3)$$

The energy stored in 1 amu is calculated as

$$E = 1.66 \times 10^{-24} (3 \times 10^{10})^2 = 1.49 \times 10^{-3} \text{ erg} \qquad (6.4)$$

In nuclear science, energy values are expressed conveniently in electron volts (eV), the energy acquired by an electron in passing through a potential of 1 volt. For nuclear energy changes on the order of millions of electron volts (1 MeV $= 10^6$ eV), the energy of 1 amu is 931.4 MeV. For the transformation of an electron (kinetic energy $= 0$; potential energy (rest mass) $= 0.0005486$ amu) into a photon (rest mass $= 0$), the kinetic energy would be

$$E = 0.005486 \text{ amu} \times 931.4 \text{ Mev/amu} = 0.51 \text{ MeV} \qquad (6.5)$$

6.15   Binding Energy

Nuclear binding energy is the "glue" that holds protons and neutrons together in the atomic nucleus. The very existence of the helium atom nucleus, consisting of two protons and two neutrons, raises a question about the classical physics requirement that two like-charged bodies repel each other inversely by the square of their distance apart. Thus, why doesn't Coulomb repulsion in the tightly bound nucleus force the breakup of the helium nucleus into two deuterium nuclei by

$$^4\text{He}^{++} \Rightarrow {}^2\text{H}^+ + {}^2\text{H}^+ \tag{6.6}$$

But helium does exist. It is one of the more stable isotopes on earth, and the inverse reaction

$$^2\text{H}^+ + {}^2\text{H}^+ \Rightarrow {}^4\text{He}^{++} \tag{6.7}$$

is an example of a thermonuclear reaction that releases nuclear energy in the form of binding energy. Thus, it is necessary to postulate the existence of *nuclear forces,* which must meet two stringent requirements:

1. They must be strong enough to overcome the coulomb repulsion forces of the many positively charged protons in the nucleus.
2. They must be very short-ranged; otherwise small stable nuclei would not exist.

The concept of nuclear force as binding energy is illustrated in Figure 6-2.

The radius of the binding-energy *well* can be approximated from the assumption of a sphere of tightly packed nucleons shown in Figure 6-1 as

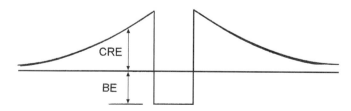

Figure 6-2. Engineering model of binding energy holding an atomic nucleus together.

$$R = R_0 A^{1/3} \tag{6.8}$$

where R_0 is a constant. Values of R_0 obtained by several independent experimental methods range from 1.2 to 1.6×10^{-13} cm; an average value of 1.4×10^{-13} cm is useful for model purposes.

Binding energy (BE) for atomic nuclei (^{A}Z) is the loss in mass of isotopic composition resulting from fusion of the component constituents (Z hydrogen atoms + N neutrons) by

$$BE = -Q = -\Delta m\ c^2 =$$
$$-\{[Z\ m(^1H) + (A\text{-}Z)\ m(n)] - m[^AZ]\}\ c^2 \tag{6.9}$$

From the following list of atomic masses (in amu), the binding energy of deuterium and helium-4 can be calculated:

Particle	Atomic Mass
n	1.008665
^{1}H	1.007825
^{2}H	2.014102
^{4}He	4.002604

For deuterium $\Delta m = 1.007825 + 1.008665 - 2.014102$

$= 0.002388$ amu

$BE = 0.002388 \times 931.4 = 2.224$ MeV

For helium-4 $\Delta m = 2(1.007825) + 2(1.008665) - 4.002604$

$= 0.030378$ amu

$BE = 0.030378 \times 931.4 = 28.29$ MeV

An index of relative stability of nuclides is given by the average binding energy per nucleon (BE/A) in the nucleus. From the examples above, ^{4}He has an average binding energy per nucleon of $28.29/4 = 7.07$ MeV/nucleon and ^{2}H has an average binding energy per nucleon of 1.11 MeV/nucleon. Figure 6-3 shows the average binding energy per nucleon as a function of mass number of stable nuclei with isotopic abundance greater than 90%. The figure shows in a qualitative way the potential for thermonuclear fusion (lighter nuclei) and nuclear fission (heavier nuclei) in relation to the peak value at a mass number of about 75 (e.g., arsenic, ^{75}As, with binding energy per nucleon of 8.4 MeV/nucleon). The binding energy of ^{235}U is 7.6 MeV/nucleon.

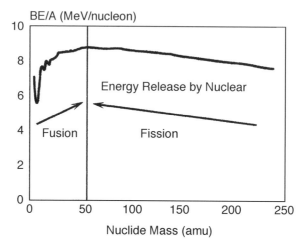

Figure 6-3. Average binding energy per nucleon as a function of the mass number of stable nuclides with isotopic abundance greater than 90%.

6.16 Nuclear Stability

Nuclear stability involves the ability of atoms to last "forever." Many isotopes of the elements have "excess" energy in their nuclei and are unstable over time. Radioactive decay is the process of spontaneous disintegration to release excess energy from the nucleus. Spontaneous disintegration requires that the released "decay" energy be positive (Q > 0, an exoergic process) and therefore that Δm be negative. The process is given by the equation

$$A \rightarrow B + b \tag{6.10}$$

where A is the radioactive nuclide, B is the resulting nuclide, and b is the emitted radiation.

Three types of radioactive decay are important in the science of nuclear energy. They are described by conditions of the excess energy of the radioactive nuclide:

1. Size (the nucleus is too large)
2. Wrong nuclear composition (too many or too few neutrons per proton)
3. Excited state (deexcitation with emission of the excess energy).

The rate of radioactive decay is a property of each radioactive nuclide, which is observed by measurement and recorded in tables of the nuclides. The value is given by the radionuclide's half-life, a measure of the time needed for an initial amount of a radionuclide to decay to half the original amount. Half-lives range from 10^{-21} seconds to 10^{16} years or more. Some natural environmental radioactive nuclides are ^{3}H (tritium), with a half-life of 12.26 years; ^{14}C (radiocarbon), with a half-life of 5730 years (useful for dating Egyptian mummies); and ^{40}K (f = 0.012%), with a half-life of 1.25×10^9 years.

6.17 Types of Radioactive Decay

Alpha Particle Decay   All elements with $Z > 83$ are radioactive because the nuclei reach a size at which the internal coulomb repulsion forces become large with respect to the nuclear forces. The common mode of decay of these elements is alpha particle emission, in which the radionuclide decays by

$$^A Z \Rightarrow {}^{A-4}(Z\text{-}2) + {}^4\text{He} + Q \tag{6.11}$$

For example, for decay of radium to radon,

$$^{226}\text{Ra} \Rightarrow {}^{222}\text{Rn} + \alpha^{++} + (4.8 \text{ MeV}) \tag{6.12}$$

After emission, the alpha particle slows down by multiple collisions and eventually picks up two electrons to form a ^{4}He atom.

Beta Particle Decay   As the number of protons in the nucleus increases in the periodic table, the number of neutrons needed to satisfy nuclear force balance (neutron/proton ratio) broadens, but if additional neutrons are added to (or taken away from) the natural N/P ratio nuclides, the resulting nuclei become unstable with respect to beta decay. Beta decay is the spontaneous emission of a high-energy electron from the nucleus in which a neutron changes into a proton by

$$^A Z \Rightarrow {}^A(Z+1) + \beta^- + \nu + Q \tag{6.13}$$

For example, for natural radiocarbon,

$$^{14}\text{C} \Rightarrow {}^{14}\text{N} + \beta^- + \nu + Q \ (156 \text{ keV}) \tag{6.14}$$

The beta particle slows down and becomes an "ordinary" electron. The antineutrino (ν) is required to conserve momentum. The carbon atom, with six protons in the nucleus, has become a nitrogen atom, with seven protons in the nucleus.

If the radionuclide has too few neutrons, beta decay takes place with the emission of a positively-charged electron (positron), which on slowing down annihilates an ordinary electron, producing two photons of 0.51 MeV (the rest mass of electrons by Eq. 6.5). An example is the radioactive isotope of sodium with one fewer neutron than stable ^{23}Na (Z = 11) decaying to neon (Z = 10) by

$$^{22}\text{Na} \Rightarrow {}^{22}\text{Ne} + \beta^+ + \nu + \text{Q} \ (2.84 \text{ MeV}) \qquad (6.15)$$

In essence, the positron may be considered an antielectron that borrows 1.02 MeV of rest mass energy and gives it back on the annihilation to two photons. For this to occur, Q must be greater than 1.02 MeV.

Gamma-Ray Decay In many cases, radioactive isotopes exist with excess energy and deactivate by releasing one or more photons of energy from the nucleus without a change in nuclear composition. The excited state of the atomic nucleus is called a metastable state, with the mass number followed by the letter m. The process is given by

$$^{Am}\text{Z} \Rightarrow {}^{A}\text{Z} + \gamma \ (\text{Q}) \qquad (6.16)$$

The radioisotope technetium-99m, which is used as a gamma-radiation source for medical diagnosis and therapy, decays by

$$^{99m}\text{Tc} \Rightarrow {}^{99}\text{Tc} + \gamma \ (\text{Q} = 0.14 \text{ MeV}) \qquad (6.17)$$

6.18 Properties of Radionuclides

Two major properties of radioactive isotopes of elements are (1) the mode of decay and (2) the kinetics of the decay process. The modes are given by the *decay scheme,* which describes how the radionuclide decays (how much of what radiations are emitted with what energies). The kinetics describe the rate of decay, which is characterized by the *half-life* of the radionuclide.

The moderately simple decay scheme of the radionuclide potassium-42 (^{42}K) is illustrated in Figure 6-4. This decay scheme provides the information that for every 100 disintegrations of ^{42}K nuclei, there will be emitted 19 beta particles with (up to) 1.99 MeV of energy (the rest

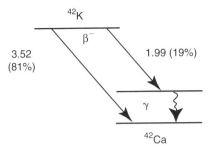

Figure 6-4. Decay scheme of the radioactive isotope potassium-42 (^{42}K).

of the beta decay energy is carried away by the antineutrino), 81 beta particles with (up to) 3.52 MeV of energy, and 19 gamma rays with 1.53 MeV of energy to form 100 atoms of calcium-42 (^{42}Ca).

The process of radioactive decay follows a first-order disintegration process in which the rate of decay (D) is proportional to the amount present, as given by

$$D = -dN/dt = \lambda N \qquad (6.18)$$

where λ = the decay constant (units of $1/t$) characteristic of each radionuclide.

The solution of this first-order differential equation is

$$N = N_0\, e^{-\lambda t} \qquad \text{or} \qquad D = D_0\, e^{-\lambda t} \qquad (6.19)$$

where N_0 = the number of atoms (D_0 = number/second decaying) at time $t = 0$.

The characteristic half-life, $T_{1/2}$, is tabulated for the radionuclides as the time for one-half of the initial number to decay: $N \Rightarrow N/2$. Thus, the half-life is given by $T_{1/2} = \ln 2/\lambda$. The units of radioactive decay originally were measured in Curies, based on the radioactivity of 1 gram of radium as 3.7×10^{10} disintegrations per second (dps). Currently, in the metric system, the unit is the becquerel, which is defined as a rate of 1 dps.

6.2 BASIC ELEMENTS OF NUCLEAR POWER

Nuclear power suggests the use of nuclear energy in large quantities. This originally was employed for military might, and later was used for civil applications to generate electricity on a large scale. Indeed, the discovery of uranium fission in 1938 opened the nuclear power age with the building of the first experimental nuclear reactor in 1942 and the first use of an atom bomb in 1945. The ensuing development of the civilian nuclear reactor industry for the generation of electricity coincided with the military development of the hydrogen bomb, which was based on the thermonuclear fusion of hydrogen isotopes into helium. The potential for these two processes was shown in Figure 6-3. The present-day utilization matrix for nuclear energy is shown in Figure 6-5.

The three axes are the divides between nuclear fission and nuclear fusion, civil and military applications, and the time element of the nuclear processes. The United States (and other nuclear-capable countries) has abandoned the technology for peaceful applications of nuclear explosives. The breakthrough needed to acquire controlled thermonuclear reactors (to produce solar energy on earth) has not occurred. Thermonuclear fusion technology is not yet available for the human quest for abundant energy.

6.21 Nuclear Fission

The adsorption of a "thermal" neutron by a uranium-235 nucleus produces a uranium-236 nucleus that, in addition to being unstable with

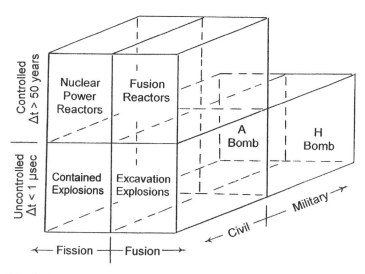

Figure 6-5. A three-dimensional matrix sketch of the utilization of nuclear energy.

respect to alpha particle decay, can split into two approximately half-size fission products, which then fly apart by coulomb repulsion. The nuclear-fission reaction is given by

$$^{235}U + n_{th} \Rightarrow {}^{A1}Z_1 + {}^{236-(A1+\nu)}Z_{92-Z1} + \nu n + (Q \cong 200 \text{ MeV})$$

$$(6.20)$$

Thermal neutrons are neutrons that have lost their high kinetic energy through collisions in the reactor with energies of low thermal temperature ($\sim 1/40$ eV). The amount of energy released by the fission is approximately 200 MeV. What makes the process useful as a large-scale energy source is the release of the ν high-energy neutrons, which then allow further fissions in the fuel. If more that one of the ν neutrons results in further fission, the rate of fission grows exponentially, and the system becomes an atom bomb explosive in less than 1 microsecond. If the system is engineered to allow just enough neutrons (at critical mass of fuel) for fission to continue at a constant rate, the system becomes a nuclear power reactor that could operate for more than 60 years. Technical improvements in current and newly designed nuclear reactors could extend the lifetime of the nuclear fission era until thermonuclear fusion is fully developed.

6.22 Available Energy from Uranium Fuel

Nuclear fuel is obtained from deposits of uranium in ore scattered throughout the world. The main chemical form of the uranium in the ore is U_3O_8. One ton of U_3O_8 produces 0.85 ton of uranium, which in nature consists primarily of ^{238}U and has only 0.71% ^{235}U. The 0.85 ton of uranium contains 0.0059 ton of ^{235}U. This amount of ^{235}U contains 1.38×10^{25} atoms of ^{235}U, which in a nuclear reactor could produce (at ~ 180 MeV prompt energy per fission) about 40 million kWh of electricity. The remainder of the ~ 200 MeV per fission resides in the radioactive fission products, which have great potential value as high-specific-energy sources.

The advantage of nuclear power over chemical fuel power is the much greater specific energy: the amount of energy released per unit mass of fuel. For comparison, 1 ton of high-quality coal (with heating value of 12,000 Btu/lb) can produce about 2800 kWh of electricity. Thus, on a specific energy basis, one ton of U_3O_8 is equivalent in electricity production to $\sim 14,000$ tons of coal.

The magnitude of uranium fuel resources for the generation of electricity comes in two sizes, as determined by both technology and political decisions. The technology size follows the methods discussed in Chapter 3. It concerns our ability to measure how much uranium ore exists in the earth's crust at attainable depths. Current information on uranium resources [1] uses the categories (by confidence level) of

RAR = reasonably assured resources, known with current technology and within a given cost range

EAR = estimated additional resources, inferred from known deposits with available exploration data

EAR-I: based on direct geologic data

EAR-II: based on indirect geologic data

SR = speculative resources, based on indirect evidence and geologic extrapolation

In the model of the supply-demand relationship, the IAEA [1] forecast of demand and market-based production (in kilotonnes) through 2050 in the middle-demand scenario is summarized in Table 6-1.

The reserves-production (R/P) ratio as a function of market price for world reserves of uranium was calculated for the categories of higher confidence level as follows:

Reserves

RAR + EAR-I at $80/kg U = 3.107 Mt U

RAR + EAR-I at $130/kg U = 3.93 Mt U

Production (2000)(at 80% recovery) = 28.1/0.8 = ∼ 35.1 kt U

The resulting R/P ratios were

Table 6-1   Forecast supply-demand relationship for uranium fuel resources [1]

Year	Demand (kt U)	Market-Based Production (kt U)
2000	61.6	28.1
2025	95.0	81.9
2050	177.0	159.6

$$R/P = 3.107/0.035 = \sim 89 \text{ years at } \$80/kg \text{ U}$$
$$R/P = 3.93/0.035 = \sim 112 \text{ years at } \$130/kg \text{ U}$$

The magnitude of uranium reserves available for the generation of electricity also is based on political decisions about how the uranium will be used to generate electricity. These decisions involve the choice of adapting a *once-through* fuel cycle (the *open* fuel cycle) or a *reprocessing* fuel cycle (the *closed* fuel cycle). The United States adapted the open fuel cycle after the decision by the Carter administration (1977–1981) to limit commercial nuclear energy in the United States to a *once-through* fuel cycle because of fears of international terrorism involving the theft of enriched uranium and the plutonium produced in the fuel in the *reprocessing* fuel cycle. The technical aspects of these fuel cycles are discussed in Section 6.23.

6.23 Nuclear Power Reactors

During the early development of nuclear reactor technology, several designs were considered for commercial utilization, among them the light-water reactor, the heavy-water reactor, the high-temperature gas-cooled reactor, and the liquid-metal-cooled reactor. These designs are described in most textbooks on nuclear technology (e.g., [2, 3, 4]). At the beginning of the twenty-first century, there were more than 100 nuclear reactors operating in the United States and more than 440 in the world, with an average electric power capacity of about 840 MWe. Table 6-2 shows the main types of nuclear power plants operating (and under construction) in the world from data from the International Atomic Energy Agency (IAEA).

Table 6-2 Nuclear power plants in the world by reactor type[a]

Reactor Type	Operational		Under Construction	
	Number	Capacity (GWe)	Number	Capacity (GWe)
Pressurized water (PWR)	267	241.2	11	10.4
Boiling water (BWR)	92	82.0	4	5.0
CANDU (PHWR)	40	20.5	7	2.6
Gas-cooled (GCR)	22	10.7	0	0.0
Light-water graphite (LWGR)	16	11.4	1	0.9
Fast breeder (FBR)	3	1.0	1	0.5
Total	440	366.8	24	19.4

[a] As of June 2005, compiled from the IAEA database on the Internet (2005).

The most numerous of these nuclear reactors (almost all in the United States) are light-water-cooled reactors, in which ordinary water is used as the coolant to carry the nuclear-produced heat in the reactor vessel to the electric turbine-generator and as the moderator to slow down the fast-fission neutrons to thermal-energy neutrons. The light-water reactor is produced in two types: boiling-water and pressurized-water reactors [5]. The engineering differences among the current reactor designs are shown in the schematics in Figure 6-6.

The boiling-water and pressurized-water reactor designs of the light-water nuclear reactors are the most common. In a boiling-water reactor, the coolant water is allowed to boil in the reactor vessel and the steam flows from the reactor vessel to the turbine-generator. In a pressurized-water reactor, the coolant water is pressurized to prevent boiling in the reactor vessel and the superheated liquid water is conveyed to a heat exchanger within the reactor vessel where steam is produced to turn the turbine-generator.

The other reactor types listed in Table 6-2 are the heavy-water reactor (HPWR) (termed the CANDU reactor for Canadian-deuterium-uranium), which uses natural uranium (UO_2) as the fuel and heavy water (HDO and D_2O) as the coolant and moderator, the gas-cooled reactor (GCR), which uses gases (e.g., helium and CO_2) as the coolant and graphite as the moderator; the Russian-design light-water reactor (RBMK), which uses water as the coolant and graphite as the moderator; and the fast breeder reactor (FBR), which uses a liquid metal (e.g., sodium) as the coolant and requires no moderator [fission of enriched uranium and production of plutonium (Pu) by neutron capture is done with fast neutrons].

Utilization of nuclear energy for electric power generation is developing worldwide in many nations. A summary [6] of the worldwide distribution of nuclear power facilities is given in Table 6-3. The table is divided in two parts, with the first part listing the 10 countries with the greatest reliance on nuclear-generated electricity supply and the second part listing the 10 countries with the largest installed power capacity.

6.24   The Light-Water Uranium Fuel Cycle

Figure 6-7 shows a *box model* of the full light-water reactor fuel cycle. The front end (manufacture of fuel rods for the nuclear reactor vessel) is essentially the same for the *open* and *closed* cycles. The back end (following discharge of the *spent* fuel rods) results in the *closed* cycle.

A typical PWR

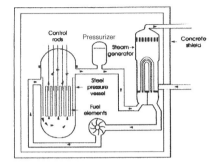

A typical PHWR (Candu)

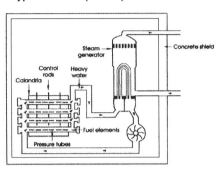

A typical BWR

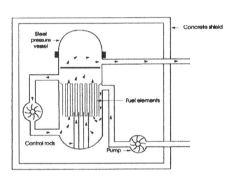

A typical Gas Cooled Reactor (Magnox)

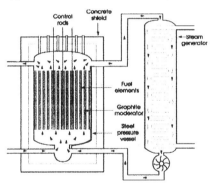

A typical LWGR (RMBK)

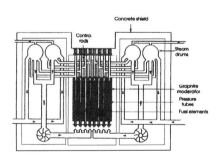

A typical Advanced Gas Reactor (AGR)

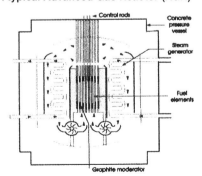

Figure 6-6. Schematics of currently operating nuclear reactors. (Source: World Nuclear Association).

Table 6-3   (a) Nuclear-generated electricity by share of electricity[a]

Country	Nuclear-Generated Electricity (TWh)	Share of Total Electricity (%)	Number of Reactors	Installed Capacity (GWe)
1. Lithuania	14.3	79.9	2	2.370
2. France	420.7	77.7	59	63.473
3. Slovakia	17.9	57.4	6	2.472
4. Belgium	44.6	55.4	7	5.728
5. Sweden	65.5	49.6	11	9.429
6. Ukraine	76.7	45.9	13	11.268
7. Slovenia	5.0	40.5	1	0.676
8. South Korea	123.3	40.0	19	15.880
9. Switzerland	25.9	39.7	5	3.220
10. Bulgaria	16.0	37.9	4	2.722
11–31. Rest of world	1714.8	n/a	314	245.897
Total	2524.7	16.0	441	363.135

(b) Nuclear-generated electricity by installed power capacity[a]

Country	Installed Capacity (GWe)	Nuclear-Generated Electricity (TWh)	Share of Electricity (%)	Total Number of Reactors
1. United States	97.485	765.7	19.9	103
2. France	63.473	420.7	77.7	59
3. Japan	45.521	230.8	25.0	54
4. Russia	20.793	138.4	16.5	30
5. Germany	20.643	157.4	28.1	18
6. South Korea	15.880	123.3	40.0	19
7. Canada	12.080	70.3	12.5	17
8. United Kingdom	12.048	85.3	23.7	27
9. China	11.471	79.0	[b]	15
10. Ukraine	11.268	76.7	45.9	13
11–31. Rest of world	52.473	379.1	n/a	86
Total	363.135	2524.7	16.0	441

[a] As of May 30, 2004, listing the 10 countries (a) with the largest share of nuclear-generated electricity and (b) the largest installed power capacity.
[b] Nine reactors on the Chinese mainland and six reactors in Taiwan.
Source: World Nuclear Association, *Pocket Guide 2004*, August 2004.

The *open* fuel cycle ends with the arrow from Interim Storage to Permanent Storage. The *closed* fuel cycle continues with the arrow to Fuel Reprocessing. If the present generation sends the spent fuel rods to "permanent" storage, it is likely that future generations will "mine" those rods for needed energy, probably at great cost.

The full cycle consists of the following operations:

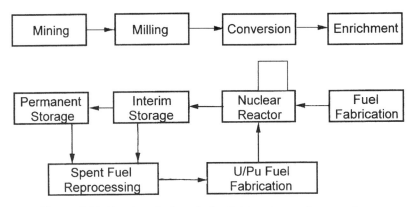

Figure 6-7. Uranium fuel cycle for light-water nuclear reactors.

Step in Cycle	Fuel Form
Uranium ore mining and milling	U_3O_8 "yellowcake"
Chemical conversion	$U_3O_8 \Rightarrow UF_6$ (volatile)
Enrichment	0.7% $^{235}U \Rightarrow 2\text{--}3\%$ ^{235}U
Fuel fabrication	$UF_6 \Rightarrow UO_2$ (in high-T clad rods)
Burnup	Amount of fission of ^{235}U in the fuel rods
Reprocessing	Recycling of U, Pu, and fission products
Nuclear waste disposal	Nonuse of high-specific-energy sources

6.25 Generation IV Nuclear Reactors

The reactors described in Section 6.23 are termed Generation II reactors, which have incorporated engineering lessons learned from the original generation of nuclear reactor designs. Research is in progress internationally on the design of advanced reactors (termed Generation III reactors) for near-term (2010–2030) use. In addition, since 2000, a group of 10 countries have formed a Generation IV International Forum (GIF) to look ahead to the period after 2030 for further development of nuclear reactor designs that are secure, safe, and cost-effective to meet the expected increase in electricity demand on a sustainable basis. A review [7] of the status of the GIF effort is at the World Nuclear Association website: www.world-nuclear.org/.

Table 6-4, from that review, shows the main characteristics of the six systems chosen for detailed design. The reactor type is determined by the coolant system that provides the listed temperature range. The neutron energy ranges from fast (as emitted in the fission event), to epithermal (partially slowed down), to thermal (as used in Gen II light-water reactors). Four of the six designs operate at temperatures sufficiently high for potential use of the reactors for thermochemical

Table 6-4   Survey of generation IV nuclear reactor designs [7]

Reactor Type	Coolant	Neutron Energy	Temperature (°C)	Size(s) (MWe)	Potential Uses
Gas-cooled	Helium	Fast	850	288	E + H
Lead-cooled	Pb-Bi	Fast	550–800	300–400; 1200	E + H
Molten-salt	Fluoride salts	Epithermal	700–800	1000	E + H
Sodium-cooled	Sodium	Fast	550	150–500, 500–1500	E
Supercritical	Water	Either	Very high	1500	E
Very-high temperature gas-cooled	Helium	Thermal	High	250	E + H

dissociation of water to produce hydrogen (H) (see Chapter 8) as well as electricity (E).

6.26   Nuclear Safety

Nuclear reactor safety has been a controversial subject throughout the over 50-year history of nuclear power reactors operating around the world. Major fears cited by activists against the use of nuclear energy for generation of electricity are as follows:

1. The reactor will explode like an atom bomb.
2. A loss of coolant accident (LOCA) with a simultaneous failure of the emergency core cooling system (ECCS) will lead to a melt-down of the reactor.
3. A massive release of radioactive materials under accident conditions.
4. Continuous release of radioactive materials under normal operating conditions.
5. Waste cooling heat will be ecologically dangerous.
6. Transportation of radioactive materials will be dangerous.
7. Plutonium (Pu) and enriched uranium fuel will be diverted to terrorists and rogue nations.
8. Radioactive waste disposal will fail and be ecologically dangerous.

These fears continue in spite of the reliability and safety regulation efforts of the International Atomic Energy Agency (IAEA) for the world and the U.S. Department of Energy (DOE) [enforced by the U.S. Nuclear Regulatory Commission (NRC)] for the United States. The policies for safety include

1. Evaluation of all possible types and severity of accidents
2. Provision of engineered safeguards with oversight of construction and operation to prevent accidents
3. Evaluation of dependable means of containment and site security
4. Continuous inspection of designs, construction, operation, licensing, and compliance
5. Conducting of supporting safety research and testing programs

6.27 Nuclear Waste

Nuclear waste (the world's greatest oxymoron?) has been a subject of technical and social research for more than 50 years. The fear of using reprocessed nuclear fuel and high-specific-energy radioactive fission products may be considered *irrational* by future generations. The dictionary defines an *oxymoron* as "a combination of contradictory or incongruous words." It defines *nuclear* as "relating to or utilizing atomic energy" and *waste* as "discarded as worthless, defective, or of no use." To consider radioactive materials (atomic energy) containing a million fold greater specific energy than chemical fuels as waste materials will indeed be considered a great folly in the future, when fossil fuel energy resources become scarce and renewable energy resources are unable to meet the energy demand of an ever-increasing world population.

The amount of nuclear fuel consumed in a nuclear reactor during an operating cycle is termed *burnup* and is expressed in units of megawatt-days per metric ton (MWd/Mt). For a typical 1000-MWe light-water reactor containing 100 Mt of fuel and operating at 33.3% (thermal to electric energy) conversion efficiency for 3 years (1100 days), the burnup would be 3000 (MWth) $\times$ 1100 (days)/100 (Mt) = 33,000 MWd/Mt. This burnup would consume 3.3 Mt of ^{235}U, or 3.3% of the fuel in the reactor core loading. The spent fuel, "no longer useful to society," would contain 96.7 Mt of expensively processed nuclear fuel, about 10 to 12 kg of fissionable plutonium, and about 3.3 Mt of high-energy radioactive fission products.

What are we to do with the spent (waste) fuel? There are three options:

1. Today (for the past 50-plus years) it is kept at nuclear reactor sites to "delay and decay."
2. Tomorrow (perhaps by 2012) it will be stored in temporary or permanent disposal sites such as the Yucca Mountain federal repository in New Mexico.
3. Some time in the future it will be reprocessed to extract the spent fuel and high-energy radioactive fission products for use in a recycle economy.

6.3   THE OKLO NATURAL NUCLEAR REACTORS ON EARTH

In September 1972, uranium was found in the ore deposit at Oklo in Gabon, Africa, with an abnormal isotopic composition that led to speculation that spontaneous fission chain reactions had occurred on earth in the remote past. By international agreement, such observations are reported to the International Atomic Energy Agency. Those investigations led to the realization that about 2 billion years ago natural nuclear reactions occurred on earth (at least at Oklo) for some 500,000 years.

Two major scientific questions needed resolution:

1. How could natural nuclear fission have occurred some 2 billion years ago?
2. How could it be determined in 1975 that it indeed happened?

Those questions were evaluated in two international scientific meetings of the IAEA [8, 9], as noted in the references cited [3, 5, 6, 7] at the end of this chapter. Several Internet sources are available for further information on the OKLO nuclear reactors.

The first question involved nuclear reactor technology for enrichment, criticality, moderation, and poisons [10]. For enrichment, since the half-life of ^{235}U is less than that of ^{238}U, the 0.71% isotopic fraction of ^{235}U in uranium 2 billion years ago was in the light-water-reactor range of more than 2%. For criticality, sufficiently concentrated uranium salts had been deposited in the riverbeds to constitute a critical mass for sustained nuclear fission. For moderation, the water acted as the moderator used in today's nuclear reactors to moderate the neutrons (i.e., reduce the energy of the emitted high-energy neutrons by collisions without absorbing them for capture by nearby ^{235}U nuclei). For poisons, other nuclides that can absorb neutrons faster than does uranium, such as cadmium, which is used for control rods in modern nuclear reactors to control the rate of fission, were absent in the uranium ores. Thus, these four conditions necessary for sustained fission reactions were met for the estimated duration of 500,000 years of natural nuclear fission at Oklo.

The second question involved the change in isotopic composition of the uranium in Oklo ores, the geologic permanence of the natural reactor site, and the measurement of the stable (nonradioactive) end products of the fission product decay chains. The very first indications that the isotopic fraction of ^{235}U was much lower than 0.71% gave credence to the possibility that consumption of ^{235}U could have taken place. Extensive geologic investigation of the site showed that the uranium ore beds had not undergone deep subsidence with time, enabling discovery of the ore deposit. The enrichment in the fission end product isotopes by several beta decays from radioactive fission products in relation to the "normal" isotopic composition of those elements was readily measured by many scientists around the world.

A third scientific question may be asked—What does it all mean today?—in relation to both the evolution of natural history and its importance for modern nuclear waste disposal. For scholars of natural history, the phenomenon of a natural release of some 100 billion kWh of nuclear energy in one small place when the earth was about half as old as it is today must have been of ecological importance. For the administration of nuclear waste disposal priorities, the state of preservation of the Oklo "fossil nuclear reactor" site as it is today must lend some credence to the ability of geologic formations to *store* nuclear waste for billions of years.

6.4 THERMONUCLEAR FUSION

This chapter closes with a short look at the potential for true solar energy on earth: duplication of the sun's method for producing energy continuously by thermonuclear reactions with isotopes of hydrogen. The basis for this possibility, similar to Einstein's equation for the equivalence of mass and energy, is the Boltzmann equation, which relates energy (E, in eV) to temperature. (T, in degrees Kelvin):

$$E = kT \tag{6.21}$$

where k = the Boltzmann constant
k = 8.62×10^{-5} eV/°K

At room temperature T = 298°K, gaseous hydrogen atoms (H_2) have a statistical average thermal energy of ≈ 1/40 eV. At solar temperature T = 15×10^{6}°K. Thus, the mean energy of solar atoms is ≈ 1.3 keV.

The overall source of the sun's energy is the fusion of heavy hydrogen nuclei (deuterium, ^2H) by

$$^2H + {}^2H \Rightarrow {}^4He + BE(\gamma) \tag{6.22}$$

to form helium-4 nuclei with release of the binding energy as gamma radiation. This reaction is extremely difficult to produce on earth. The more promising reaction, which has been achieved on earth in the form of the hydrogen bomb and lasts less than 1 microsecond, is between the hydrogen isotopes deuterium and tritium:

$$^2H + {}^3H \Rightarrow {}^4He + {}^1n + Q(17.6 \text{ MeV}) \tag{6.23}$$

The technical breakthrough needed to lengthen the lifetime of the "thermonuclear fusion reactor" from less than 1 microsecond to more than 50 years involves the ignition temperature of the process. The rate of thermonuclear reactions rises rapidly with temperature above a critical (ignition) temperature (T_c) that corresponds to a minimum threshold energy.

For the deuterium-deuterium reaction, the energy required for ignition is about 20 keV, corresponding to a critical temperature of T_c ~ 10^8 °K. For the deuterium-tritium reaction, the critical temperature is T_c ~ 10 keV, which is much easier to attain on earth though still a difficult technical problem for sustained energy production. Therefore, if a mixture of deuterium and tritium is heated to T > ~ 10 keV, the rate of reactions will increase and

If the mass is held together as dE/dt $\Rightarrow \infty$, an explosive hydrogen bomb results.

If the process is controlled as dE/dt $\Rightarrow 0$, a thermonuclear fusion power reactor results.

The scientific breakthrough needed is finding a way to achieve controlled thermonuclear reactions. Deuterium at 10^7°K is a plasma, $^2H \Rightarrow d^+ + e^-$; the deuterium plasma, stripped of the electrons, can be kept in a magnetic "bottle." The possible control technologies are as follows:

1. Magnetic fusion: Keep T at exactly T_c (hard to do).
2. Laser fusion: Limit the d^+ concentration in the "reactor" (easier to do).
3. Allow "miniexplosions" in the "reactor" (hard to do).

Research to date is approaching the magnetic fusion criteria of the theoretical threshold of the deuterium concentration-energy-time product (nvt) $> 10^{14}$ ion-seconds. Today's magnetic containment equipment is capable of reaching (nvt) greater than 4×10^{12} ion-seconds. One can hope that humanity will reach the ultimate energy source—true solar energy on earth—before nuclear fission fuels run out.

6.5 SUMMARY

The chapter provided a succinct review of nuclear energy with an *engineering model* of the atomic nucleus, which consists of protons and neutrons and is surrounded by orbiting electrons. The model is sufficient to understand the concept of nuclear binding energy as it relates to the twentieth-century discovery of nuclear energy in the forms of nuclear fission and nuclear fusion. Nuclear energy is pictured here only for its civil application as an energy source in the same manner that a hammer can be pictured as a useful tool that can pound nails into walls without noting that a hammer also can crack skulls. The chapter described the basics of nuclear power reactors, with observations on nuclear safety and nuclear waste, and concluded with a discussion of the prehistoric occurrence of natural nuclear reactors in Africa and the future goal of thermonuclear fusion as Solar Energy on Earth in addition to our Solar Energy from the Sun.

REFERENCES

[1] International Atomic Energy Agency. *Analysis of Uranium Supply to 2050.* STI/PUB/1104. Vienna: IAEA, 2001.
[2] R. A. Knief, *Nuclear Energy Technology.* New York: McGraw-Hill, 1981.
[3] R. L. Garwin and G. Charpak, *Megawatts and Megatons.* Chicago: University of Chicago Press, 2002.

[4] S. W. Heaberlin, *A Case for Nuclear-Generated Electricity.* Columbus, OH: Battelle Press, 2004.

[5] World Nuclear Association, *World Power Reactors.* Information series, London WNA, September 2004.

[6] World Nuclear Association, *WNA Pocket Guide-2004.* August 2004.

[7] World Nuclear Association, *Generation IV Nuclear Reactors.* Information Series, London WNA, April 2005.

[8] International Atomic Energy Agency, *The OKLO Phenomenon,* STI/PUB/405. Vienna: IAEA, 1975.

[9] International Atomic Energy Agency, *Natural Fission Reactors,* STI/PUB/475. Vienna: IAEA, 1978.

[10] G. A. Cowan, "A Natural Fission Reactor." *Scientific American* 235: 36, 1976.

7

RENEWABLE
ENERGY RESOURCES

7.0 RENEWABLE ENERGY

Renewable energy has had many definitions over the last 50 years as various constituencies have focused on different social causes and technical aspects. A broad definition from the environmental movement since the 1960s includes any energy source that is "alternative" to "conventional" fossil (and, for some, nuclear) fuels. This definition includes geothermal energy, which is not really a renewable resource in that it requires hundreds to thousands of years to replace the heat extracted from the geothermal deposit.

7.01 Types of Renewable Energy

It is convenient to classify primary energy resources into three major types:

1. *Primordial:* These are the heat sources that originated with the formation of the earth. Since early times, boiling water at hot springs was used for cooking. Today, geothermal resources (hydrothermal and petrothermal) are deposits of heat close enough to the earth's surface to be extractable in commercial quantities through geothermal wells.
2. *Fossil:* This is decayed organic (carbon-containing) matter fossilized over millions of years that can be extracted by the mineral extraction industries.

3. *Renewable:* This consists of the energy fluxes available on a daily basis from incoming solar and lunar sources.

Energy available for the human quest also can be classified by origin as solar, lunar, and terrestrial. From the diagram of energy flow on earth (Figure 2-1), the power obtained is as follows:

Solar	Thermal (heat and radiation)	76 PW
	Hydropower (the hydrologic cycle)	40 PW
	Kinetic (wind power)	0.37 PW
	Biomass (for food and power)	0.04 PW
Lunar	Tidal (coastal water wave power)	3 TW
Terrestrial	Geothermal heat (estimated resource)	400 EJ

The important renewable resources that are examined in separate sections in this chapter are

Hydroelectric power	Section 7.1
Solar thermal and photovoltaic power	Section 7.2
Wind power	Section 7.3
Biomass energy (biofuel)	Section 7.4

7.02   Consumption of Renewable Energy

The data on trends in energy consumption in the United States since 1900 were reviewed in Section 2.31. A key observation there was the steady decline in the growth rate as the magnitude of total energy consumption increased. During the last 40 years societal awareness of *appropriate technology* has resulted in increased societal pressure for a change to environmentally clean, green energy resources. The extent to which this desire has been achieved during the last five years is given in Table 7-1. The data compiled by the EIA include geothermal energy as a renewable. The data for 2002 were listed as preliminary.

The data for those five years show several interesting trends, including the decline of four of the five renewable energy types as the total consumption of energy in the United States slowly rose, the large growth rate of wind energy from a small initial level, the relatively greater decline of hydroelectric energy as the major form of renewable energy, and the relatively high negative growth rate of renewable energy as a percentage of total U.S. energy consumption. DOE/EIA noted that the slight decline in the fraction of renewable energy consumption from

Table 7-1   Renewable energy (quads) in U.S. energy consumption 1998–2002 [1]

Energy Type	Year					m.a.g.r. (%/a)
	1998	1999	2000	2001	2002	
Hydroelectric	3.30	3.27	2.81	2.20	2.67	−8.19
Solar	0.07	0.07	0.07	0.07	0.06	−2.39
Wind	0.03	0.05	0.06	0.07	0.11	28.5
Biomass	2.82	2.87	2.89	2.66	2.74	−1.37
Geothermal	0.33	0.33	0.32	0.31	0.30	−2.14
Total renewable	6.55	6.59	6.15	5.31	5.88	−4.31
Total U.S.	95.14	96.76	98.93	96.31	97.55	0.45
Percent renewable	6.88	6.81	6.21	5.51	6.03	−4.76

7% in 1998 to 6% in 2002 reflected the continued high cost of their technologies compared with the cost of competing technologies, especially those using natural gas.

Most of the renewable energy facilities installed today are for electricity generation, especially hydroelectric and geothermal facilities; thus, the proportion of renewable energy consumed for generating electricity in relation to the total renewable energy consumed each year during this five-year period remained nearly constant at approximately 66%. The data from DOE/EIA [1] for electricity generation (in TWh) from renewable energy are listed in Table 7-2 for comparison. Indeed, the growth rates of the three renewable resources for electricity generation (hydroelectric, wind, and geothermal) are about the same. The difference lies in the use of solar and biomass for thermal power, but both of those uses are relatively small. Solar power is developing as a preferred source of energy in the residential sector, and biomass in the form of wood is used for space heating and increasingly as a raw

Table 7-2   Electricity generation (TWh) in the United States from renewable energy 1998–2002 [1]

Energy Type	Year					m.a.g.r. (%/a)
	1998	1999	2000	2001	2002	
Hydroelectric	323.3	319.5	275.6	217.0	263.6	−7.95
Solar	0.5	0.5	0.5	0.5	0.5	2.53
Wind	3.0	4.5	5.6	6.7	10.5	29.0
Biomass	58.8	59.6	60.7	57.0	59.4	−0.25
Geothermal	14.8	14.8	14.1	13.7	13.4	−2.78
Total renewable	400.4	399.0	356.5	294.9	347.5	−5.86

material in the transportation sector for the production of ethanol as an additive to gasoline fuel to increase combustion efficiency and reduce tailpipe emissions.

The DOE/EIA forecast of growth of renewable energy consumption in the United States, which is shown in Table 7-3, indicates a larger mean annual growth rate of about 2.1%/a compared with the negative growth rate of −4.3%/a over the period 1998–2002. The assumption was that social demand and improved technology will combine to make renewable energy more competitive over the next 25-year period and that its use for electric energy generation will grow at a greater rate than it will for thermal energy applications.

7.1 HYDROELECTRIC POWER

Hydroenergy is the energy in moving (falling) water. Waterwheels have been in use for millennia to grind grain and distribute irrigation water. The major use of large-scale hydropower today is for the generation of electricity. The National Hydropower Association (see http://www.hydro.org/ for "Facts You Should Know about Hydropower") notes that hydropower is a clean, renewable, and reliable energy source that serves national environmental and energy policy objectives. The association also notes that total U.S. hydroelectric power capacity is more than 100 GW (including pumped storage capacity), which is two-thirds of the 148 GW installed and ultimate capacity estimated by Hubbert [3] in the 1960s. The installed capacity in 2000 [1] was 79.4 GW and was forecast [2] to be 78.7 GW in 2025.

Large-scale hydropower is produced at hydroelectric power stations by allowing water to fall at a controlled flow rate from a reservoir,

Table 7-3   Forecast of renewable energy consumption through 2025 [2]

Year	Primary Energy			Electric Energy		
	Renewable (Quad)	Total (Quad)	Fraction (%)	Renewable (PWh)	Total (PWh)	Fraction (%)
2000	5.96	72.2	8.3	0.316	3.46	9.1
2005	6.71	74.6	9.0	0.378	3.68	10.3
2015	7.75	82.3	9.4	0.405	4.51	9.0
2025	8.78	89.3	9.8	0.429	5.31	8.1
m.a.g.r. (%/a)	2.1	0.9		2.1	1.9	

created by building a dam on a river, through a "penstock" (pipe) to a turbine-generator some distance below. Figure 7-1 shows a schematic of a typical hydropower station.

The electric power generated is governed by three key parameters in the design of the station:

1 The head: the vertical distance (h, in meters) of water flow fall
2. The flow: the volumetric flow rate of the water (Q, in m^3/s)
3. The conversion factor, F, of the turbine-generator equipment

by the potential energy (PE, in joules), which is given by

$$PE = m\ g\ h \qquad (7.1)$$

for a falling stream of water. The power produced (P, in kW) is given by

$$P = F\ (\rho Q)\ g\ h \qquad (7.2)$$

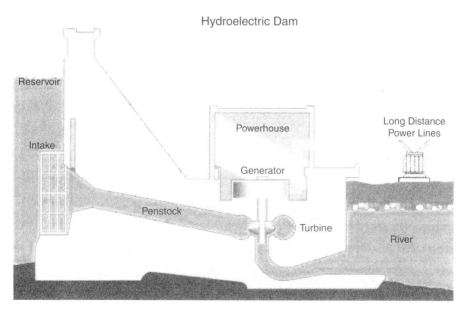

Hydroelectric Dam

Reservoir

Intake

Powerhouse

Long Distance
Power Lines

Generator

Penstock

Turbine

River

Figure 7-1. Cross section of a typical hydroelectric power station [4].

where ρ = density of the water (kg/m³) [with ρQ = mass flow (m, in kg/s)]

    g = force of gravity (m/s²)

Hydroelectric power stations are classified by head into two types:

1. High-head: h ≥ 20–1000 m, e.g., Hoover Dam in Nevada
2. Low-head: h ≤ 5–20 m

High-head facilities can store water at an intermediate head during high-flow periods and use it during dry periods to level out the power generation rate partially. Low-head plants cannot economically store water; thus, their output varies with seasonal rainfall.

Hydropower resources in the United States are under the jurisdiction of the Federal Energy Regulatory Commission (FERC). A review of the hydroelectric generating capacity was compiled by FERC in 1992 [5] and is shown in Figure 7-2.

Hydropower is especially important in the western states. A major system is in place on the Columbia River in the northwest. The large Bonneville station generates more than 1 million kW, and with the other stations on the river, the total power generation is in excess of 8.4 million kW. The forecast by DOE/EIA that very little additional power capacity from hydropower is expected over the next 20 years in the United States makes hydroelectricity an important but not exciting resource in the future. From an environmental point of view, hydropower is considered one of the more environmentally clean energy resources, but the concept of *renewable* is being challenged in that silting in the upper and lower reservoirs may limit station lifetime to 50 to 200 years.

7.2 SOLAR ENERGY

The sun provides the energy of life on Earth. As a continuously operating fusion reactor with an interior temperature of several million degrees Kelvin, the sun radiates energy throughout the solar system. The relationship between radiation energy flux (emissivity, ϵ) and temperature is given by the Stefan-Boltzmann equation:

$$\epsilon = \sigma\, T^4 \tag{7.3}$$

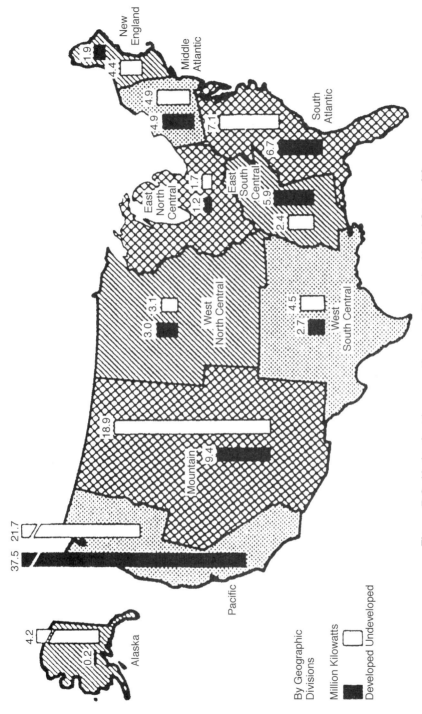

Figure 7-2. Hydroelectric generating capacity in the United States [5].

143

where ϵ = emissivity from the surface (in W/m^2)
σ = Stefan-Boltzmann constant (5.67×10^{-8} W/m^2 K^4)
T = radiative temperature (Kelvin, raised to the *fourth* power).

The energy flux radiates isotropically (as $1/R^2$) from the sun's surface at a temperature of about 6000°K and reaches the earth (at R = 1.5×10^8 km) at a mean surface temperature of 288°K (15°C). This is the temperature that allows flora and fauna to exist on earth (particularly comfortably in the temperate zones).

7.21 The Solar Constant

Solar energy reaching the earth appears to have been constant over the human time period and is spread out in wavelength. The spectral solar irradiance on the earth as a function of wavelength, (λ, in μm) is shown in Figure 7-3.

The irradiance (the power flux per unit wavelength, in W/m^2-μm) peaks at the visible light wavelength of 0.5 μm. The integral of the spectrum, the area under the curve, is the *solar constant,* which is equal to 1.35 kW/m^2. The solar spectrum can be divided into three major wavelength regions with their respective abundance as follows:

Radiation	Wavelength (μm)	Abundance (%)
Ultraviolet	<0.4	9
Visible light	0.4–0.7	45
Infrared	>0.7	46

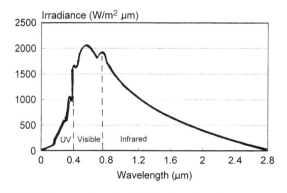

Figure 7-3. The wavelength distribution of solar power on earth.

The spectrum of short-wavelength radiation provides the *resource base* of 178 PW shown in Figure 2-1. The *albedo* (direct reflection away from the earth) accounts for about 35% of the short-wavelength energy entering the atmosphere. Thus, the *internal* resource base is about 116 PW. Only a small fraction of this power is usable.

7.22   Solar Energy "Reserves"

Some solar energy is available each day everywhere on earth. The usable forms are

Direct beam (thermal) radiation, which can be focused to a collector

Diffuse (thermal) radiation, scattered from clouds, which cannot be focused

Secondary forms converted to biomass (fuel), wind energy, and hydropower

The energy received each day per unit area on earth is the *daily insolation* (H, in units of kWh/m²-d). The amount varies by latitude, season, and weather. Values for the United States [6] range from about 1 to 7.5 kWh/m²-d, with the larger values abundant in the sunbelt southwestern states, as shown in Figure 7-4.

Except for direct heating, these daily reserves must be converted into other useful forms by some conversion technology. These uses include the following:

1. Direct thermal use such as greenhouses, crop drying, and sunbathing. The energy absorbed is

$$E \text{ (kJ/kg)} = m \, C_p \, \Delta T \qquad (7.4)$$

where m = mass (kg)
C_p = specific heat (kJ/kg-C)
ΔT = increase in temperature (C)

2. Heating and cooling with flat-plate collectors. The energy consumed is

$$E \text{ (kJ/kg)} = \epsilon \, A_p \, H \qquad (7.5)$$

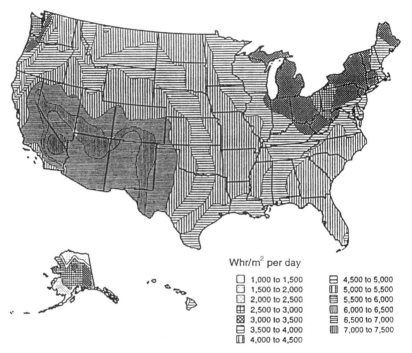

Figure 7-4. Average daily direct (beam) solar radiation in the United States [6].

where ϵ = thermal efficiency (~15 %)
     A_p = perpendicular area (m²)
     H = daily insolation (kWh/m²-d)

3. Electricity generation with photovoltaic collectors. The energy absorbed in a silicon wafer p,n semiconductor is

$$E \ (eV) = V \ \mathbf{e} \ (n\mu_n + p\mu_p) \tag{7.6}$$

where V – applied voltage (volts)
     $\mathbf{e}$ = electron charge (eV)
     $n\mu_n$ = number concentration of electrons (N/m³) in the conduction band with electron mobility μ_n (N/m²-s)
     $p\mu_p$ = number concentration of positive holes (P/m³) in the conduction band with hole mobility μ_p (P/m²-s).

7.23 Solar Electricity

Large-scale (MWe) solar electricity can be generated by two major conversion technologies: solar thermal and solar electric. The former is achieved by concentrating the energy of incoming solar radiation into a *heat carrier* and then converting the heat of the heat carrier into electricity. The latter technology involves the direct conversion of solar radiation into electricity by solid-state photovoltaic cells.

A major problem in solar electricity is the low flux of direct-beam solar radiation per unit surface area. As noted in Twidell and Weir [7], a single solar tracking bowl (even with a diameter of 30 m in a 1-kW/m^2 insolation zone) would receive at most thermal power of 700 kW(th) and generate, at a conversion efficiency of 20% or more, about 200 kW(e), which is too little for large utility networks. Thus, current research and development programs are concentrating on increasing the conversion efficiencies of the concentrating systems to reduce the size of the collecting fields for the generation of megawatt quantities of electric power.

Solar Thermal Electricity Generation facilities for solar thermal electricity have distributed collectors that are classified in three types of power systems by the means of collecting solar energy, which in the U.S. Department of Energy's Solar Energy Technologies Program [8] are termed troughs, dish engines, and power towers. The trough system consists of parabolically curved trough-shaped reflectors which focus on a pipe carrying a heat carrier such as oil to the generator. Figure 7-5 shows a trough collector system in the Mojave Desert in southern

Figure 7-5. A distributed parabolic trough collector system for the generation of electricity (source: U.S. Department of Energy).

California that was built in the 1980s. There are nine such systems there, producing electric power in amounts ranging from 14 to 80 MWe.

Dish/engine systems are collectors that use mirrors to reflect the incoming solar radiation and focus the concentrated energy flux to a small area in the thermal receiver. The receiver transfers the heat to a heat carrier fluid that becomes the working fluid for the generator. Several experimental facilities to develop these systems are under construction.

Power tower systems are central generating stations surrounded by a field of solar energy collectors that focus the reflected energy to a carrier fluid, (e.g., water to form steam) in the tower that is used as the working fluid in the power station for conversion to electricity. Figure 7-6 illustrates the power tower concept with a photograph of the Southern California Edison Company's 10-MW(e) test facility at Barstow, which produced more than 38 GWh of electric energy during its six years of operation in the 1980s. Each of the 1818 plane mirrors reflected direct radiation to the tower boiler to provide the steam to operate the system's turbine-generator.

Figure 7-6. An early (1982–1988) solar power tower facility for generation of electricity. Source: Southern California Edison Corporation.

Solar Photovoltaic Electricity The direct conversion of incoming direct-beam solar radiation into electricity, avoiding the efficiency limitations of first converting the solar energy to thermal energy and then to electricity, probably will make electricity generation a primary large-scale use of solar energy. As a low-specific-energy resource, it is well suited to low-specific-energy needs such as heating and cooling individual buildings. With the ability to provide a building with electric energy, it might be reasonable to assume that in the not too far off future, civil engineering practice will require all new construction, large and small, to be covered with photovoltaic surfaces. Figure 7-7 shows a current solar electric system for a house in California that integrates photovoltaic panels into an awning in the backyard, providing both electricity for the house and shade for the back porch.

The photovoltaic cell operates in accordance with the physics of the photoelectric effect, which is one of the ways ionizing radiation interacts with matter. In gamma-ray decay, as described in Section 6.17, the gamma-ray energy (Q) in MeV is much greater than energy spectrum of solar radiation (in eV; given by the wavelength spectrum in Figure 7-3) and interacts with surrounding atoms by the Compton scattering process. In this process, the high-energy gamma ray ejects a large number of electrons from the materials in its path as it slows down and

Figure 7-7. A photovoltaic array acting as a backyard porch cover while providing electricity to the home. Source: U.S. Department of Energy.

ejects a last electron before it disappears with zero rest mass energy. The photoelectric effect with solar radiation is like that last interaction in which an electron is removed from an atom, but in many materials, the electron is elevated in energy to an outer orbit and emits that energy as a photon.

In semiconductor materials such as silicon, in which the atoms are in a crystal lattice structure, electrons can be freed from the bound state (to their atoms) by radiation with energies in eV across an energy gap (for silica, the energy gap is about 1.1 eV) to a conduction band where they can move under an applied voltage. In a photovoltaic cell, a device that allows the absorption of solar radiation to result in an electric current, the applied voltage is provided by "doping" adjacent layers with n- and p-impurities that result in an excess of electrons in an n-type semiconductor and an excess of missing electrons ("holes") in a p-type semiconductor. The current across the photovoltaic cell results from the excess electrons in the n-type layer moving across the boundary to the p-type layer. A sketch of n-type and p-type semiconductors in a photovoltaic cell and the location of the cell in a photovoltaic panel is shown in Figure 7-8.

The energy absorbed in a silicon wafer p,n semiconductor was given in Eq. 7.6. The electricity produced is determined by the size and number of photovoltaic cells in the installation. The conversion efficiency (in percent) is given as

$$\epsilon = 100 \times \text{electricity output/solar energy input} \qquad (7.7)$$

where the electric energy output for the array is measured in watt-hours (Wh). The fraction of solar energy absorbed varies with the engineering design of the photovoltaic collector array and the daily insolation (in Wh/m^2-d). Efficiencies of commercial arrays range from about 10% to 20%.

7.3 WIND ENERGY

Generation of electricity with wind energy is achieved by atmospheric wind power rotating a rotor-blade propeller on a wind tower rotator shaft that turns a wind turbine. Wind energy can be provided with a single wind tower or a field of towers that constitute a wind farm. Modern wind towers, such as the one pictured in Figure 7-9a range in

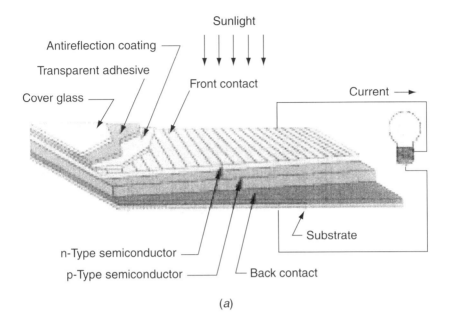

Sunlight

Antireflection coating

Transparent adhesive

Cover glass

Front contact

Current ⟶

n-Type semiconductor

p-Type semiconductor

Back contact

Substrate

(a)

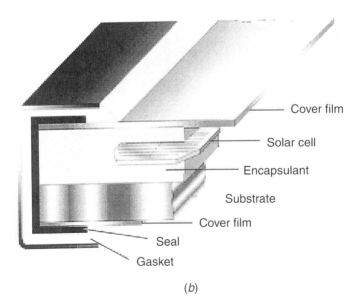

Cover film

Solar cell

Encapsulant

Substrate

Cover film

Seal

Gasket

(b)

Figure 7-8. Schematic of a photovoltaic panel (upper) sketch of a p,n semiconductor arrangement and (lower) its location in the photovoltaic panel.

(a)

(b)

Main Shaft   Brake

Gear-box

Transmission   Generator

Hub

Housing   High Speed Shaft

Blades

Tower

Figure 7-9. (a) A modern wind tower. (b) Schematic of the turbine assembly [9].

capacity from 500 kW to 2 MW on towers that are 40 to 60 m in height and with rotor diameters from 40 to 75 m. The major components of the wind turbine system are shown schematically in Figure 7-9b.

7.31 Wind Power Rate

Wind energy produced in a wind turbine is a function of several parameters; key among them are

Air density (ρ, kg/m^3)
Rotor-blade-sweep area (A, m^2)
Wind velocity (V, m/sec)
Time (t, sec) and a conversion efficiency factor

The instant energy (kWs) produced is the power rate (P, kW) multiplied by the instant time period (sec), as given by

$$dE = P \, dt \tag{7.7}$$

The kinetic energy generated in the wind turbine is given by

$$KE = \tfrac{1}{2} \, m \, V^2 \tag{7.8}$$

and the mass flow rate of air, m (kg/s), is given by

$$m = \rho \, A \, V \tag{7.9}$$

Thus, the general equation for the power rate generated in the wind turbine is

$$P = \tfrac{1}{2} \, (\rho A V) \, V^2 = \tfrac{1}{2} \, \rho \, A \, V^3 \tag{7.10}$$

At a given site, the value of ρ for the prevailing barometric and temperature conditions is

$$\rho = \rho_0 \, (288B/760T) \tag{7.11}$$

where ρ_0 = density of dry air at standard temperature and pressure
    = 1.225 kg/m^3 at 288°K (15°C, the mean surface temperature of the earth) and 760 mm Hg

B = barometric pressure (mm Hg)
T = ambient air temperature at rotor height (K)

7.32   Wind Turbine Conversion Efficiency

The power rate P is the total energy per unit time in the airstream, of which only a fraction can be converted to electric energy in a wind turbine. The theoretical limit, which was estimated around 1920, is calculated to be about 59%. Modern wind turbines are designed to achieve a conversion efficiency of about 40%.

A second major limit to the power rate is the large and continuous variation in wind speed. This variation is especially important since rate is proportional to the third power of the wind velocity. A twofold change in velocity results in a eightfold change in power output (see Eq. 7.10). Thus, a major problem in estimating the annual output of a wind turbine (in kWh) is the time variation in mean wind velocity over an annual period. Long-term wind statistics for a proposed wind-farm site generally are compiled as a Weibull probability function (see Figure 7-10) to estimate the mean number of hours per year at each velocity interval. The data are plotted in the form of a Rayleigh wind speed frequency distribution (see Figure 7-11).

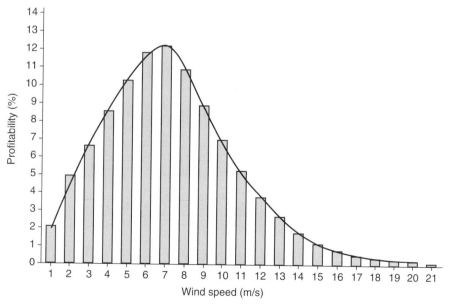

Figure 7-10. Weibull probability function for wind speed (histogram in 1-m/s intervals) [10].

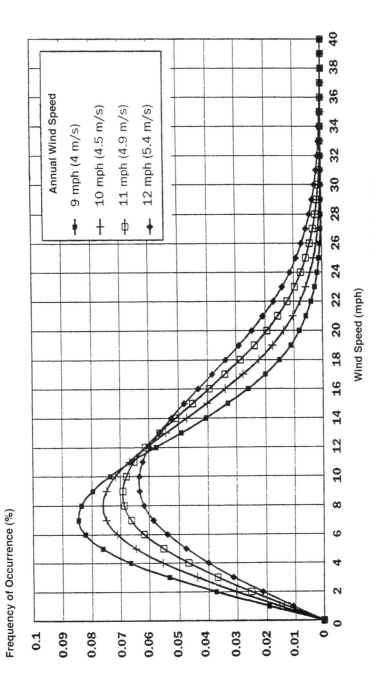

Figure 7-11. Rayleigh wind speed frequency distribution [11].

7.33   The Wind Energy Resource

Estimation of the potential for wind energy on a global basis requires an estimation of the mean wind speed over habitable landmasses of the earth's surface. These data are available as wind atlases in a general form for many populated regions of a nation. In 1994, the World Energy Council [12] estimated that about 27% of the land surface has wind speeds greater than 5.1 m/s at 10 m above the surface. In 1997, it was noted [10] that only 4% of the area might be available for generating electricity in wind farms because of unsuitable terrain, urban areas, crop cultivation, and other land uses. For an assumed generating capacity of 8 MW/km² and a capacity factor of 23%, the global potential for wind turbine power production was estimated to be 20 PWh per year. The worldwide consumption of electricity in 2001 was estimated at 13.9 PWh.

Wind power capacity is growing in the United States, reaching a value of 4,685 MW in 2002 [13], which constituted about 2% of the renewable energy that provided about 7% (263 TWh) of the 3.5 PWh of electricity consumed in the United States. The distribution of wind power capacity [13] is shown in Figure 7-12.

7.34   Estimated Cost of Wind Power

The economic viability of expanding the use of large-scale wind farms for significant electricity-generating capacity depends on the reduction of wind turbine cost ($/installed kW) and the cost of the electricity (¢/kWh) compared with the costs for competing fuels (e.g., fossil, nuclear, and solar). Construction costs of current large-scale wind turbines (e.g., 1500 kW with 77-m rotor diameter is about $1000/kW, compared with natural gas installations at about $600/kW) [14]. In 2001, a formula for estimating annual energy output based on averages of existing data was reported [15] as

$$E = 8760\, P\, (0.087V - P/D^2) \tag{7.12}$$

where P = rated power capacity (kW)
     V = mean annual wind velocity (m/s) at rotor height (~50 m)
     D = rotor diameter (m)
  8760 = hours/year

For a 1500-kW turbine with a 77-m rotor and a mean wind speed of 7 to 7.5 m/s, the estimated annual electricity generated [15] would

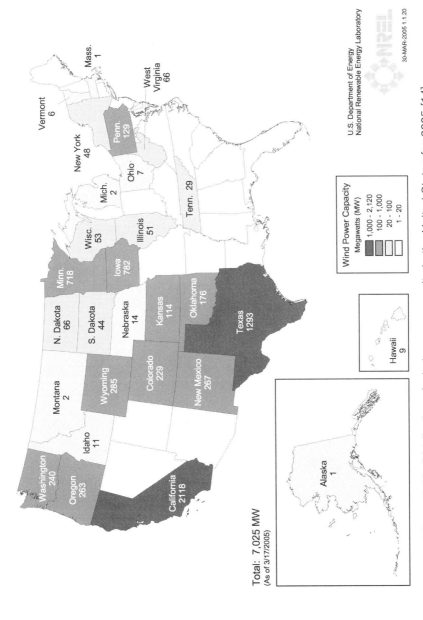

Figure 7-12. Distribution of wind power capacity in the United States for 2005 [14].

Total: 7,025 MW
(As of 3/17/2005)

Wind Power Capacity
Megawatts (MW)
1,000 - 2,120
100 - 1,000
20 - 100
1 - 20

U.S. Department of Energy
National Renewable Energy Laboratory

30-MAR-2005 1.1.20

Washington 240
Oregon 263
California 2118
Idaho 11
Montana 2
Wyoming 285
Colorado 229
New Mexico 267
N. Dakota 66
S. Dakota 44
Nebraska 14
Kansas 114
Oklahoma 176
Texas 1293
Minn. 718
Iowa 782
Wisc. 53
Illinois 51
Mich. 2
Ohio 7
Tenn. 29
New York 48
Penn. 129
Vermont 6
Mass. 1
West Virginia 66

Hawaii 9

Alaska 1

be 4.7 to 5.2 GWh/yr. The electricity cost would be between 3 and 4 ¢/kWh.

The environmental costs of wind energy on a large-scale basis are undetermined. A key driving force for wind energy is the relatively benign environmental impacts of wind turbine technology. The "fuel" is free and contains no carbon; the turbines pass through the ambient concentration of CO_2 in the atmosphere without change. However, potential negative aspects of wind energy include

Avian interaction with wind turbines

Visual impact of wind farm turbine density

Wind turbine noise

Electromagnetic interference

Land use impact of the wind power system

Safety considerations from blade throw, ice fall, tower failure, and electrical fires

It may be assumed that as the magnitude and severity of these impacts are recognized, technical improvements will ensue.

7.4   BIOMASS ENERGY

Biomass (and the bioenergy obtained from it) comes from agriculture. It is one of the oldest energy sources on earth and may become one of the most significant large-scale energy sources in the future. Biomass originates from the photosynthesis portion of the solar energy distribution and includes all plant life (terrestrial and marine), all subsequent species in the food chain, and eventually all organic wastes. Biomass resources come in a large variety of wood forms, crop forms, and waste forms. The basic characteristic of biomass is its chemical composition in such forms as sugar, starch, cellulose, hemicellulose, lignin, resins, and tannins. Many excellent overviews of biomass as an energy resource are available (e.g., [16]). The U.S. Department of Energy provides a description of the Energy Efficiency and Renewable Energy Section's Biomass Program on the Internet [17].

Bioenergy may be defined as the energy extracted from biomass for conversion into a useful form for commercial heat, electricity, and transportation fuel applications. Two basic characteristics of biomass are its heat value and its moisture content. The heat value generally is expressed in kJ/kg or Btu/lb for small amounts and in GJ/t(metric) or Btu/t(short) for commercial-size amounts. The moisture content (as a percent) of the biomass determines the more useful processes for converting it to bioenergy.

The heat value of biomass varies with the chemical composition of the organic components and their relative weight in the biomass. For example, the dry-weight fraction (in percent) for three major biomass components in trees and grasses are 40 to 60% for cellulose, 20 to 40% for hemicellulose, and 10 to 25% for lignin. The average heat value for wood, as noted in Figure 1-4 and 1-5, is about 15 MJ/kg to 16 MJ/kg. The moisture content of biomass generally varies from about 10% to 60%; it is measured by weighing a sample before and after drying. The moisture content (in percent) is expressed in two ways: wet basis and dry basis. The moisture content wet basis (MCWB) is given by

$$MCWB = 100 \times (\text{initial weight} - \text{dry weight})/\text{initial weight} \quad (7.13)$$

For bioenergy application, it is more common to express the moisture content dry basis (MCDB), which is given by

$$MCDB = 100 \times (\text{initial weight} - \text{dry weight})/\text{dry weight} \quad (7.14)$$

7.41   The Solar Biomass Resource

Bioenergy potential, although large, accounts for only a small fraction of the solar energy component resulting in photosynthesis. For growing crops for energy in temperate climates with solar radiant energy per hectare (10,000 m^2, 2.47 acres) on agricultural land averaging about 2 to 5 kWh/m^2 per day (see Figure 7-4 for the United States), an average yield of about 11 metric tons (dry)/ha-yr (about 5 U.S. tons/acre-yr) containing 15 GJ/t, the net available energy is about 165 GJ (about 0.5% of the incident solar energy)

Bioenergy resources are classified in the DOE-EERE Biomass Program in two major *platforms* that generally are distinguished as entailing bioconversion and thermal conversion processes. A sketch of the biomass-to-biorefinery program (Figure 7-13) shows the position of the two platforms in the total system.

The sugar platform includes the bulk of plant materials, such as energy crops (terrestrial and aquatic) and agricultural crop residues,

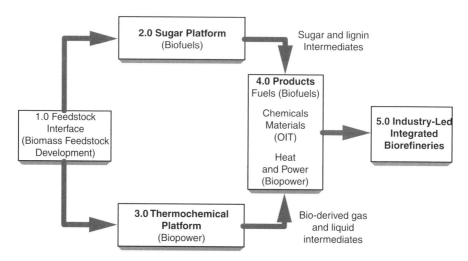

Figure 7-13. Schematic of the DOE-EERE Biomass Program [18].

which contain cellulose, hemicellulose, and lignin that can be broken down into their component sugars. Those sugars can be processed biologically to produce ethanol and other potential fuels.

The thermochemical platform includes biomass such as tree wood and postharvest and processing remains such as forest residues, forest plantations of energy crops, and municipal green wastes (such as paper products and urban tree pruning) that can be converted by thermal and chemical processes to produce an array of bioenergy fuels for thermal and electric power generation.

Both platforms require handling of large quantities of biomass materials generally distributed over large agricultural and forest regions and thus require a well-developed infrastructure to make the supply chain (harvesting, collection, storage, transport, and processing) economically and environmentally viable. Synergistic savings in effort, cost, and pollution can be achieved, such as reduction in landfill fields, air and water pollution abatement, and additions to the regional energy supply with less transportation of imported fuels.

The output from the two biomass platforms, shown as "Products" in Figure 7-13, are the feedstocks for the biorefineries that produce the bioenergy products by with a suitable choice among the many conversion processes available.

7.42 Biomass Conversion Processes

Biomass conversion processes generally are grouped into the two categories of biochemical conversion and thermochemical conversion.

Figure 7-14 shows some of the processes and forms of application that make bioenergy collectively an important part of renewable energy resources.

Large-scale processing of biomass by multiple conversion processes most likely will be concentrated in large plants called integrated biorefineries by the DOE. A schematic of the total system is shown in Figure 7-15. The central column shows the sugar platform technology, the left column shows the thermochemical platform technology, the dashed line follows the flow of internal heat and power demand, and the large arrows indicate the output products. The figure provides an excellent means to appreciate the potential for large-scale production of biomass as a useful resource for energy and commercial materials.

7.43   Environmental Aspects of Bioenergy Fuels

Widespread adaptation of biomass as an alternative energy source for large-scale applications with a power supply greater than about 100 kW will require that the cost of producing bioenergy fuels compete not only with that of fossil fuels but also with that of other renewable energy resources. In addition, the life-cycle emission of global greenhouse gases and urban air pollutants must be competitive. For example, the U.S. Clean Air Act Amendments legislation since the 1990s requires the use of oxygenated fuels and reformulated gasoline when air pollution conditions are above U.S. Environmental Protection Agency (EPA) standards. The EERE Biomass Program noted the EPA report [19] showing that biodiesel and e-diesel (ethanol blended with petro-

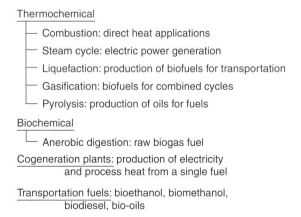

Thermochemical
— Combustion: direct heat applications
— Steam cycle: electric power generation
— Liquefaction: production of biofuels for transportation
— Gasification: biofuels for combined cycles
— Pyrolysis: production of oils for fuels

Biochemical
— Anerobic digestion: raw biogas fuel

Cogeneration plants: production of electricity
and process heat from a single fuel

Transportation fuels: bioethanol, biomethanol,
biodiesel, bio-oils

Figure 7-14. Outline of available processes for the conversion of biomass for applications of bioenergy, biopower, and biofuels.

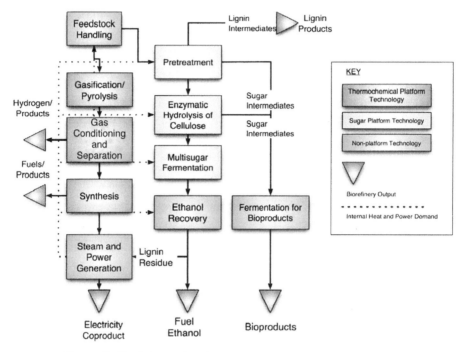

Figure 7-15. Schematic of an integrated biorefinery (from [18]).

leum diesel) fuels substantially reduced emissions of key air pollutants (except nitrogen oxides) from heavy-duty highway engines, as summarized in Table 7-4.

The environmental benefits of biofuels for the production of thermal and electric power are numerous. They include reduction in sulfur dioxide emission, since most forms of biomass contain only small

Table 7-4 Average emission impacts of biodiesel for heavy-duty highway engines[a]

Pollutant	*Reduction (in percent) for Biodiesel Blends*	
	100%	20%
Carbon monoxide	−47	−12
Hydrocarbons	−67	−20
Particulates	−48	−12
Nitrogen oxides	+10	+2

[a] Adapted from Figure ES-A, EPA [19] reported by U.S. DOE-EERE.

amounts of sulfur. Blending of biomass with coal, which may contain up to 5% sulfur, reduces a power plant's emission of SO_2 markedly compared with a coal-only operation.

The reduction in carbon emission is also significant, since the emission of CO_2 in the fuel cycle is offset by the CO_2 recycled into new biomass production. Other benefits arise from using municipal biomass wastes for bioenergy production, avoiding its decay and odor generation in landfills.

7.5   OTHER RENEWABLE RESOURCES

Two other renewable energy resources are included in this review of alternative energy resources: tidal energy (which should be called ocean energy and would include the contribution of the earth's moon as tidal energy) and geothermal energy (the contribution from the earth's interior other than the fissionable elements, thorium and uranium).

7.51   Tidal Energy

With oceans covering two-thirds of the earth's surface and undergoing twice-daily variation in sea level, the potential for harvesting energy from them has been a subject of great interest and visionary research. Tidal energy shows good potential for electric energy generation.

The ability to extract energy from the world's oceans comes in three forms:

1. Wave energy: The kinetic energy resulting from the moon's (in addition to the sun's smaller) gravitational pull on the oceans under the earth's rotation, producing a diurnal tidal effect, has the potential to utilize the kinetic energy of waves hitting continental shores.
2. Current energy: The heat cycle of tropical solar energy affects the oceans during the earth's rotation and generates kinetic energy that could be used directly to turn submerged turbine generators.
3. Thermal energy: The temperature gradient of the ocean with depth, results in a temperature difference just large enough over reasonable depths to extract thermal energy at low efficiency.

Wave energy against coastlines appears to be the more likely means of achieving useful generation of electricity. A brief explanation of the basic astronomical factors that produce tides and tidal currents is available at http://www.co-ops.nos.noaa.gov/restles1.html. A review of the engineering means to harness the tidal energy is given at http://reslab.com.au/resfile/tidal/text.html. The technology is similar to that of hydropower, with the falling head measured by the difference in height of the low and high tides. However, a big difference in technology is due to the one-dimensional fall of river flow, whereas tidal flow is "in and out" from the shoreline. This difference requires turbines that can function with flow in both directions. Tidal power facilities are best suited to large bays with narrow inlets that allow a damlike structure called a barrage to control the tidal flow through gates and turbines.

At present, only one commercial tidal energy facility is in operation (since 1966), a 240-MW power station at La Rance estuary in northern France. There are several experimental projects under way around the world; large-scale commercial exploitation may not develop until well into the future.

7.52 Geothermal Energy

Geothermal energy is literally the heat of the earth. It has been evident on the earth's surface in the forms of volcanoes, geysers, fumaroles, and hot springs since the time of the very earliest human inhabitants. A review of geothermal resources, extraction, and utilization in 1975 [20] noted that thermal energy is available everywhere in the upper 10 km of the earth's crust, with a mean temperature gradient of 20 to 30°C/km depth and a mean emissive heat flux of about 1.5 μcal/cm^2-sec. Although a very large resource base exists in the upper 10 km, the thermal concentration of the rock (kJ/m^2) is too diffuse to be an exploitable resource on a worldwide basis. Geothermal resources suitable for commercial utilization may be defined as localized geologic deposits of heat concentrated at attainable depths, in confined volumes, and at temperatures sufficient (T > ~180°C) for electric energy and (T > ~100°C) for thermal energy utilization. Nature has provided four types of thermal deposits (resources), which are listed below in order of ease of extraction:

1. Hydrothermal: trapped steam or hot water contained in geologic strata reachable by drilling that can be extracted to drive a turbine generator. It is essentially the only resource type in production today and occurs in two levels of quality:

a. Vapor-dominated: extracted mainly as dry steam run directly to the turbine
b. Liquid-dominated: extracted as hot water and flashed to steam on pressure reduction or maintained under pressure through a heat exchanger to vaporize a low-boiling-point organic fluid
2. Petrothermal: high-temperature dry rock formations [termed hot dry rock (HDR) in much of the world] that can be fracture-stimulated to allow sufficient permeability for heat-exchange extraction in the formation with cold surface water circulated in a closed loop to conversion equipment on the surface.
3. Geopressured: sedimentary formations that contain high-pressure hot water and natural gas (methane) that can be extracted by drilling. Much of the value of the resource lies in the value of the natural gas.
4. Magma: bodies of lava that have risen to reachable depths (<20 km) for heat exchange with circulating heat-carrier fluids.

The technology for conversion of geothermal fluids into electric energy is the same as that for fossil fuels; the main difference is in the properties of the working fluids. Geothermal steam, as it comes from the ground, contains gaseous impurities, such as hydrogen sulfide and radon, that generally are not permitted to be released to the atmosphere. Furthermore, the steam temperature is much lower than it is in steam from fossil fuel boilers. Instead of approximately 35 to 40% conversion efficiency in turbines, geothermal steam conversion efficiencies are on the order of 12 to 15%, and about three times the steam fluid flow is required. The more abundant lower-temperature geothermal resources have an even lower conversion efficiency and may require operation in a binary cycle power plant in which the geothermal fluid is maintained under pressure to keep it from boiling and is used to boil a lower-boiling-point fluid such as isobutane to drive the turbine. In both cases the geothermal fluid is reinjected into the ground for environmental purposes as well as to maintain reservoir fluid pressure and extract additional heat from the host rock.

Geothermal energy is used primarily for electric energy generation and thermal applications such as space heating, industrial heating and drying, and public bathing. In the United States, the main interest is in the generation of electricity. Table 7-5 lists the role that geothermal energy played in the electricity supply in the year 2002 and the forecast for future generation and capacity through 2025. DOE/EIA anticipates that the fraction of electric energy supplied from hydrothermal re-

Table 7-5 Geothermal energy in the U.S. electricity supply [2]

Year	Generation			Capacity		
	Geothermal (GWh)	Renewable (GWh)	Fraction (%)	Geothermal (GW)	Renewable (GW)	Fraction (%)
2002	13.36	308.9	4.3	2.89	91.7	3.2
2005	14.23	383.2	3.7	2.90	94.2	3.1
2015	32.31	424.5	7.6	5.11	101.2	5.1
2025	46.66	464.4	10.1	6.84	110.1	6.2
m.a.g.r. (%/a)	5.6	1.8		3.8	0.8	

sources in the United States will reach 10% of the renewable energy contribution by 2025. In 2002, the 13.38 GWh of geothermal energy represented 0.4% of the total electricity generation of 3391 GWh in the United States. This value would grow to 0.9% of the total generation of 5257 GWh in 2025.

A more up-to-date review (1997) of the status of geothermal energy [21], that included the potential of the vast Hot Dry Rock (petrothermal) worlwide resources emphasized its potential impact as an environmentally sustainable energy resource. Current information on the status of geothermal energy is available on the Internet from two sites:

1. The Geothermal Resources Council at www.geothermal.org (for the United States)
2. The International Geothermal Association at iga.igg.cnr.it/index. php (for the world)

7.6 SUMMARY

This chapter examined the potential role of renewable resources in providing a significant portion of the world's energy mix in the next two generations. The renewables in current use are essentially hydropower and geothermal energy. Rapid development is under way of solar energy in the forms of solar photovoltaic electricity, wind power, and biomass technologies. The basic science and engineering of these resources indicate that each of them could contribute to the energy mix of the world. A major advantage of these energy sources is their public acceptance as green energy resources (although each of them in the total fuel cycle contributes global and local atmospheric pollution). The major disadvantage is their low specific-energy content for use in large-scale energy-intensive applications. This review of the major types of available energy sources (fossil, nuclear, and renewable) indicates that the energy mix of the future will be time-dependent but also will involve some contribution from each of them.

REFERENCES

[1] U.S. Department of Energy, Energy Information Agency, *Renewable Energy Annual.* Report No. DOE/EIA-0603(02), Washington, D.C.: DOE, November 2003.

[2] U.S. Department of Energy, Energy Information Agency, *Annual Energy Outlook 2004.* Report No. DOE/EIA-0383(04), Washington, D.C.: DOE, January 2004.

[3] M. K. Hubbert, "Energy Resources." Chapter 2 in V. J. Yannacone, Jr., *Energy Crisis: Danger and Opportunity.* St. Paul, MN: West Publishing, 1974.

[4] Tennessee Valley Authority, Hydroelectric Dam, www.tva.gov/power/hydroart.htm (2005).

[5] Federal Energy Regulatory Commission, *Hydroelectric Power Resources in the United States Discovered and Undiscovered.* Washington, DC: FERC, 1992.

[6] Office of Technology Assessment, *Renewing our Energy Future.* Washington, D.C.: Congress of the United States, 1995.

[7] J. W. Twiddle and A. D. Weir, *Renewable Energy Resources.* Cambridge, UK: Cambridge University Press, 1986.

[8] U.S. Department of Energy, Solar Energy Technologies Program: www.doe.gov/engine/content.do?BT_CODE=SOLAR, 2005.

[9] J. F. Manwell, J. G. McGowan, and A. L. Rogers, *Wind Energy Explained.* New York: John Wiley & Sons, 2002.

[10] J. F. Walker and N. Jenkins, *Wind Energy Technology.* New York: John Wiley & Sons, 1997.

[11] P. Gipe, *Wind Energy Basics.* White River Junction, VT: Chelsea Green, 1999.

[12] World Energy Council, *New Renewable Energy Resources.* London: Kogan Page, 1994.

[13] National Renewable Energy Laboratory, National Wind Technology Center, "2005 Installed Wind Power Capacity Map," private communication, 2005.

[14] J. Johnson, "Blowing Green," Chemical and Engineering News, February 24, 2002, pp. 27–30.

[15] M. Z. Jacobson and G. M. Masters, "Exploiting Wind versus Coal." *Science* 293:1438, 2001.

[16] R. E. H. Sims, *The Brilliance of Bioenergy in Business and in Practice.* London: James and James, 2002.

[17] Energy Efficiency and Renewable Energy Section: www.eere.energy. gov/biomass/index.html.

[18] U.S. Department of Energy EERE, "Biomass Program Multi-Year Technical Plan." Draft for Review. Washington, D.C.: U.S. Department of Energy, November 2003.

[19] U.S. Environmental Protection Agency, "A Comprehensive Analysis of Biodiesel Impacts on Exhaust Emissions." Draft, EPA-420-P-02-001. Washington, D.C.: EPA, October 2002.

[20] P. Kruger, "Geothermal Energy." *Annual Review of Energy,* Vol. 1. Palo Alto, CA: Annual Reviews, 1976.

[21] J. E. Mock, J. W. Tester, and P. M. Wright, "Geothermal Energy from the Earth: Its Potential Impact as an Environmentally Sustainable Resource." *Annual Review of Energy and the Environment.* 22:305–356, 1997.

8

HYDROGEN AS AN
ENERGY CARRIER

8.0 HISTORICAL PERSPECTIVE

Hydrogen is the first chemical element in the periodic table (atomic number 1). It is the most abundant element in the universe (its relative occurrence is greater than 80%) and the third most atomic abundant element on earth. Hydrogen was identified as a chemical substance by Henry Cavendish in 1776 and was named (*hydro* = "water"; *gen* = "generator") by Antoine Lavoisier in 1785 on the basis of its stoichiometric oxidation (by combustion) to water:

$$2H_2 + O_2 \Rightarrow 2H_2O \tag{8.1}$$

Hydrogen has become an important industrial material as its physical and chemical properties have been determined over the last two centuries. Hydrogen production in the United States has reached more than 3 billion cubic feet per year. Its very low density compared to air made it (almost) ideal for the balloon airship era, and its chemical reactivity made it a feedstock in a variety of industrial processes. In the twentieth century it became a rocket fuel for propulsion into space. Its potential for becoming the next world fuel (after fossil fuels) for the generation of electricity and as a transportation fuel in the twenty-first century is the subject of the last three chapters in this study of the human quest for abundant energy.

8.01   Physical Nature of Hydrogen

Elemental hydrogen does not exist in nature as single atoms; it forms a covalent diatomic molecular gas, H_2, by

$$H + H \Rightarrow H_2 + Q(BE) \tag{8.2}$$

Hydrogen a colorless, odorless, and tasteless gas. It is the lightest known chemical substance; its density at standard temperature and pressure (STP) is 0.0899 g/L compared with air at 1.2930 g/L. Its solubility in water at 20°C is less than 20 ml/L. Liquid hydrogen is difficult to produce; it boils at -252.8°C and solidifies at -259.3°C.

Hydrogen has an unusual characteristic in that gaseous hydrogen exists as a mixture of two molecular states—ortho- and para-hydrogen—that differ by the alignment of the spins of the two electrons and nuclei. At room temperature, normal hydrogen consists of 25% para form and 75% ortho form. The two forms have slightly different properties, such as energy content and melting and boiling points. These differences become important at liquid hydrogen temperatures.

Another characteristic of elemental hydrogen is the large physical difference in its three isotopic compositions. Hydrogen consists of one proton (which identifies the atomic nucleus chemically as hydrogen) and exists in nature with 0, one, or two neutrons (resulting in mass numbers 1, 2 and 3). Thus, the isotopes of hydrogen are

$A = 1$ Hydrogen (protium) ^1H $f = 99.985\%$
$A = 2$ Deuterium ^2H $f = 0.015\%$
$A = 3$ Tritium ^3H $f = 1 \times 10^{-15}\%$

Deuterium is the "heavy" form of hydrogen and, when oxidized, forms heavy water (HDO and D_2O) with mass numbers 19 and 20 in place of 18 for H_2O. Heavy water is used as a neutron moderator in some nuclear reactors [e.g., the Canadian deuterium-uranium (CANDU) reactor]. With an atomic weight ratio between hydrogen (protium) and deuterium of $1:2$, it is relatively easy to separate deuterium from natural hydrogen compared with the separation of ^{235}U from natural uranium (mostly ^{238}U). Enriched deuterium ($f > 0.015\%$) as HD (or HDO) is used as a tracer of hydrogen (or water) in many research investigations. From a biological point of view, the osmotic pressure of enriched heavy water in tissue is different from that of natural water, and in high concentrations, heavy water is toxic to humans.

Tritium is the natural radioactive isotope of hydrogen, continuously being formed by high-energy cosmic radiation interactions with nitrogen and oxygen atoms in the atmosphere. Over hundreds of years a steady-state isotopic composition ($f = 10^{-15}\%$) is reached in which the rate of production in the atmosphere is balanced by the rate of decay with its half-life $T_{1/2} = 12.26$ years by beta particle emission to stable helium-3. As the tritium is oxidized to water in the atmosphere, it is distributed throughout the hydrologic cycle.

8.02 Chemical Nature of Hydrogen

Hydrogen is too chemically reactive to exist as an element in the free state. Its ease of providing its electron to other elements makes its natural state on earth as a compound. Hydrogen combines readily with oxygen to form water (by Eq. 8.1), with nitrogen to form ammonia (NH_3), and with carbon to form organic carbon compounds such as alkanes, C_nH_{2n+2} (e.g., octane, C_8H_{18}), and carbohydrates, $C_nH_{2n}O_n$, (e.g., glucose, $C_6H_{12}O_6$). As a result, since no natural resource of molecular hydrogen exists, its production in commercial quantities involves the decomposition of readily available hydrogen-containing compounds such as water and methane.

The chemical nature of hydrogen is complex. It undergoes several types of chemical reactions that can be illustrated by its use as

1. An active reducing agent, in which a hydrogen atom gives its electron ($H \Rightarrow H^+ + e^-$) in an acid solution by an oxidation-reduction reaction to a more active metal such as iron by

$$Fe_2O_3 + 3H_2 \Rightarrow 2Fe + 3H_2O \tag{8.3}$$

2. A hydrogenation agent. in which a hydrogen molecule is added to an unsaturated organic molecule in which two double-bonded carbon atoms open the bond to two hydrogen atoms in a reaction such as

$$C_xH_yCOOH + H_2 \Rightarrow C_xH_{y+2}COOH \tag{8.4}$$

3. A bonding agent, in which hydrogen atoms form chemical bonds with other elements to form hydrides. This can occur in two ways: as a covalent hydride such as water or ammonia by

$$2\,H_2 + O_2 \Rightarrow 2H_2O \quad \text{or} \quad 3H_2 + N_2 \Rightarrow 2NH_3 \quad (8.5)$$

and as an ionic hydride, in which a hydrogen atom accepts an electron from an atom in a metal (or alloy) lattice structure and forms a metal hydride in which the hydrogen atom acts as an anion $[H \Rightarrow H^- + p(hole)]$. For example, hydrogen reacts with sodium to form the ionic hydride NaH:

$$2Na + H_2 \Rightarrow 2NaH \qquad (8.6)$$

where sodium is the cation (Na^+) and hydrogen is the anion (H^-). The hydride form of hydrogen with metal alloys, such as iron-titanium and several rare earths, is more the covalent type of bond. These alloy metal hydrides hold promise as storage tanks in automobiles since the formation process is exoergic (heat is released) and a hydrogen fuel supply can be achieved by heating the alloy to dissociate the hydride and deliver the hydrogen as needed.

8.03   Energetics of Hydrogen

The promise of hydrogen as a transportation fuel is based on its thermal properties. The combustion specific energy released by hydrogen compared with other combustion fuels was shown in Figure 1-5 in relation to the specific energy released by nuclear fuels. For the potential of hydrogen as a transportation fuel, it is more germane to compare it with the current combustion fuels gasoline and natural gas. The energetics of these three fuels are listed in Table 8-1

It is easy to see the relative advantages and disadvantages of hydrogen compared with the other two fuels from these data. The major thermal advantage of hydrogen over both gasoline and natural gas is

Table 8.1   Energetics of transportation fuels

	Hydrogen	Gasoline	Natural Gas
Specific energy			
kJ/kg	12.5×10^4	4.45×10^4	4.8×10^4
Btu/lb	52×10^3	19×10^3	20×10^3
Volumetric energy (kJ/m³)			
Free state	10.4×10^3	32.0×10^6	37.3×10^3
Gas, 2400 psi	1.51×10^6	—	6.29×10^6
Liquid	8.52×10^6	32.0×10^6	21.1×10^6

its specific energy, which is almost three times the combustion energy of the other two fuels. The major disadvantage of hydrogen is the fact that it is the lightest gas on earth, which makes its volumetric energy at atmospheric pressure 3.5 times smaller than that of gaseous natural gas and 2.5 times smaller as liquid hydrogen than that of liquid natural gas (LNG) and more than 3.5 times smaller than that of gasoline. Thus, for hydrogen to be useful in automobile fuel tank sizes, it must be compressed to very high pressures—greater than 5000 psi (350 bar)—to attain the same driving distance per refill.

8.1 HYDROGEN AND ELECTRICITY AS PARALLEL ENERGY CARRIERS

It should be clear from this discussion of the chemical nature of hydrogen that it is not a primary source of energy. However, neither is electricity, which has become a favored choice of energy under Axiom 2 (the desire for a *pleasant habitat* and a life of *comfort and ease*). It can be said that hydrogen may become a favored choice as a transportation fuel under the same axiom. Figure 8-1 shows the concept of hydrogen and electricity as parallel energy carriers, based on the relative ease of converting each into the other when needed: electricity into hydrogen by electrolysis and hydrogen into electricity by fuel cells.

8.11 Why Hydrogen?

Invoking Axiom 2, many reasons may be advanced for moving into the hydrogen fuel age as quickly as possible. The order of importance of these reasons may differ among the many communities with agen-

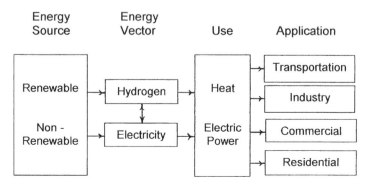

Figure 8-1. Hydrogen and electricity as parallel energy carriers.

das, but collectively they should sum to a large enough mass to convince the general public, let alone governments, to get moving.

From the energy supply and the environmental protection communities, the reasons for using hydrogen should include the following:

1. Transition to a fuel (hydrogen) that allows replacement of fossil fuels, which will increase in cost as the remaining finite supplies dwindle while the cost of hydrogen will decrease with advances in technology.

2. Transition to a fuel (hydrogen) that increases the ability to achieve wider global distribution of a sustainable energy supply for a nationally indigenous economy, especially for less-developed countries, and for smoothing of balance-of-trade inequalities among trading nations. It is well known that oil imports drain about $1billion from U.S. economy per week

3. Transition to a fuel (hydrogen) with the highest specific energy of the chemical fuels (as shown in Figure 1-5).

4. Transition to a fuel (hydrogen) prepared from a primary resource with an almost unlimited supply (water) when the technology for avoiding carbon-containing resources is achieved.

5. Transition to a fuel (hydrogen) that allows *appropriate technology* to focus the use of low-specific-energy resources (renewable energy) on low-energy-intensive applications and high-specific-energy resources (nuclear energy) on high-energy-intensive applications.

6. Transition to a fuel (hydrogen) that reverses the growth of greenhouse gas emission in the global environment and air pollutants in metropolitan air basin environments.

7. Transition to a fuel (hydrogen) that provides economic growth in a new industry over the next two generations.

The technology described in these three chapters should make these possibilities seem worth pursuing and achieving.

8.12 Competitive Uses for Hydrogen

The discussion in Section 4.2 on the ability to meet the rapidly growing demand for natural gas must be applied in a similar fashion to the possible future of a rapidly growing demand for hydrogen fuel. It has been noted that the production of hydrogen in the United States already has reached more than 3 billion cubic feet per year. However, the parameters of the discussion are not the same. Hydrogen is not a fossil fuel. If water becomes the primary source of hydrogen by any of the means to be examined soon, the concept of ultimate reserves and R/P ratios becomes meaningless. However the price of hydrogen in its useful applications does become an important parameter in choosing how to allocate production rate during the growth in demand over the buildup period of the next two generations.

Hydrogen like natural gas has competing uses as a chemical feedstock in industrial processes and as a thermal energy fuel. As was noted in Section 8.02, hydrogen is used chemically as a reducing agent in the mineral industry, as a hydrogenation agent in the petroleum industry, and as a bonding agent in the chemical industry. As a fuel, hydrogen already is used for high-temperature welding and for space exploration. Large-scale future applications are possible as a fuel for power plant fuel-cell generation of electricity, as a coolant in superconductor technology, and as a fuel in transportation applications such as motor vehicles with internal-combustion engines, motor vehicles with fuel-cell electrical engines, marine vessels, aviation jet engines, and space travel.

In analogy to Table 4-1, some of the many uncertainties about the sustainability of economic development with future uses of hydrogen are listed in Table 8-2.

8.2 THE HYDROGEN ENERGY FUEL CYCLE

The hydrogen fuel cycle is a simple one, that can be put into a four-box model:

Table 8-2 Factors that raise questions about the sustainability of hydrogen demand

Competitive uses for hydrogen
Chemical versus Energy utilization?
Potential for rapid growth of hydrogen utilization
Will consumption be limited by supply or demand?
Estimates of hydrogen resources and reserves
Is this question necessary?
Hydrogen for electric power generation
Will your grandchildren afford to heat their homes with any fuel?
Hydrogen as a parallel energy carrier with electricity
Will the concept of *"hydricity"* [1] come to pass?

1. *Production:* There are at least four general ways to produce hydrogen.
2. *Storage:* There are at least three ways to store hydrogen.
3. *Distribution:* There are at least three ways to distribute hydrogen.
4. *End Use:* There are at least two major ways to use hydrogen

8.21   Hydrogen Production

There are at least four general ways to produce hydrogen in the large quantities that will be required for a hydrogen fuel age. Chemical methods, electrolysis, thermal methods, and biological conversion processes. These methods vary in technical complexity, cost, and environmental impact, all of which have not been determined fully for large-scale production.

Chemical Methods   Early methods of hydrogen production involved oxidation-reduction reactions that used metals more reactive than hydrogen in acid solution, with the metal replacing the hydrogen stoichiometrically, the reverse of Eq. 8.3, with H_2SO_4 replacing the water. Other, more reactive metals, which are more expensive, are zinc, tin, aluminum, sodium, and potassium. Currently, the most commonly used industrial method for hydrogen production is the steam-reforming method: the removal of the oxygen from water by carbon. The earlier method was the process of reacting steam with hot carbon (coke) by the reaction

$$C + H_2O \Rightarrow H_2 + CO \tag{8.7}$$

The product mixture was called water gas and was used for heating and illuminating purposes. The current process is steam-methane reforming with natural gas by the reaction

$$CH_4 + H_2O \Rightarrow 3H_2 + CO \tag{8.8}$$

The commercial separation of the hydrogen from the carbon monoxide is then carried out by the *shift* reaction, in which the CO is oxidized to CO_2 by

$$CO + H_2O \Rightarrow H_2 + CO_2 \tag{8.9}$$

Not only is the separation of the hydrogen easier to process, the amount of hydrogen has increased by the net process

$$CH_4 + 2H_2O \Rightarrow 4H_2 + CO_2 \qquad (8.10)$$

The process is illustrated in Figure 8-2. Note that hydrogen fuel can be made either in large-scale central facilities or in small-scale (e.g., home garage) units.

8.21a A Wee-Bit of Electrochemistry

This is a good place to introduce two key aspects of electrochemistry that affect the efficiency of electrochemical devices such as electrolysers (for production of hydrogen), as described next, and fuel cells (in preparation for examining the technology of fuel cells as automotive engines in Chapter 9). The electrochemistry of water electrolysis (and fuel cells) involves the conductance of electrolytes and the temperature dependence of electromotive force (emf or voltage).

Pure water is a very poor conductor of electricity. Ionic substances (electrolytes) must be added to electrolyser cells to increase the conductivity of the solutions in the cells to useful levels. The resulting conductance of an electrolytic solution depends on both the concentration of the ions added and their characteristics (e.g., charge and mobility).

Since electrolytic solutions obey Ohm's law, the current (I, amperes) through the electrolyte is proportional to the potential difference between the electrodes (E, volts):

$$E = I R \qquad (8.11)$$

where R is the solution resistance (ohms). The conductance through an electrolyte [L, in units of siemens (S), $1 \text{ S} = 1 \text{ ohm}^{-1}$] is the reciprocal of the resistance:

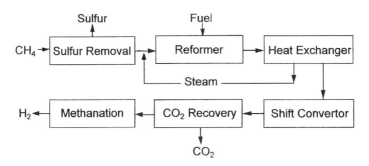

Figure 8-2. Schematic of the steam-methane reforming process.

$$L = 1/R \tag{8.12}$$

In an electrolytic cell, the specific conductance (κ, S/m) of the electrolytic solution is given by

$$\kappa = \frac{L\,l}{A} = \frac{1/R\,l}{A} = 1/kR \tag{8.13}$$

where κ = specific conductance of the electrolyte solution (S m^{-1})
   A = cross-sectional area of an electrode surface (m^2)
   l = length between electrodes (m)
   k = cell constant, the effective value of A/l for the actual cell geometry (m).

The specific conductance generally is used on a molar basis in which the molar conductivity (Λ) of an electrolytic solution of molar concentration C is expressed as

$$\Lambda = \frac{\kappa}{C} \tag{8.14}$$

The specific conductance is a function of the molar concentrations and mobilities of the cell's ionic species. For a single electrolyte in the cell (e.g., KCl, for which the specific conductance has been well measured as a function of concentration), the molar conductivity is given by

$$\Lambda = \alpha\,F\,(U_+ + U_-) \tag{8.15}$$

where α = fractional concentration of the electrolyte
   F = Faraday constant (96,485 coulomb/mol)
   U_+, U_- = ionic mobility of the cation and anion, respectively, as given by

$$U_{i.} = \frac{l}{t\,(dE/dl)} \; (m^2/Vs) \tag{8.16}$$

The temperature dependence of the electromotive force in an electrolytic cell is obtained from Eq. 8.21, in which the Gibb's free energy is the electrical work, given by

$$\Delta G = - \text{n E } F \tag{8.17}$$

where n = number of mols of electrolyte
E = electromotive force (V)
F = Faraday constant

For constant enthalpy of the cell reaction (i.e., $\Delta H = 0$), the change in Gibb's free energy is equal to the change in mechanical work, $T\Delta S$:

$$\Delta G = - T\Delta S = - \text{n E } F \tag{8.18}$$

Thus, the dependence of E on T (at constant pressure) is given by

$$\left(\frac{\partial E}{\partial T}\right)_p = \frac{\Delta S}{nF} \tag{8.19}$$

By measuring the cell voltage as a function of temperature to calculate the change in entropy (ΔS) by Eq. (8.19) and with the value of ΔG from Eq. (8.17), the total change in energy as a function of temperature can be estimated by Eq. 8.21, below.

Electrolysis   A quick surfing of the Internet searching for "electrolysis of water" yields a large number of hits, many of them high school chemistry experiments that start with the approximate wording "When a direct electric current is passed through water containing a small amount of sodium sulfate, gases are produced at the two electrodes: molecular hydrogen at one electrode and molecular oxygen at the other." College-level texts note that pure water is a very poor conductor of electricity but that the addition of an electrolyte (substances, such as H_2SO_4 and KOH, that result in conducting solutions) decreases the electrical resistance, which results in the decomposition of water by the equation

$$2H_2O \Rightarrow 2H_2 + O_2 \tag{8.20}$$

A review of several advanced methods (as of 1990) for the electrolytic production of hydrogen was prepared by Dutta [2]. The methods reviewed include (1) alkaline water electrolysis, (2) solid polymer electrolysis, and (3) high-temperature steam electrolysis.

Alkaline Water Electrolysis   The most widely used commercial technology is alkaline water electrolysis (AWE), as illustrated in Figure
8-3. This method is essentially similar to the student experiment described above.

In the commercial cell, the electrolyte is a solution of 25 to 35%
KOH operating at a temperature of about 80°C. The electrode reactions
are

$$H_2O + 2e^- \Rightarrow H_2 + 2OH^- \qquad \text{at the cathode}$$

$$2OH^- \Rightarrow \tfrac{1}{2}O_2 + 2OH^- + 2e^- \qquad \text{at the anode}$$

resulting in an overall reaction one-half of that given in Eq. 8.20.

The mean electric energy consumption in the electrolysis of water
is about 50 kWh/kg hydrogen. The cost of the electric energy makes
large-scale electrolytic production of hydrogen uneconomical compared
with the steam-methane reforming method. Work is under way to improve the AWE technology with an advanced alkaline electrolyser that
would increase cell efficiency somewhat and reduce the electricity requirement to about 43 kWh/kg. Another advance in this technology

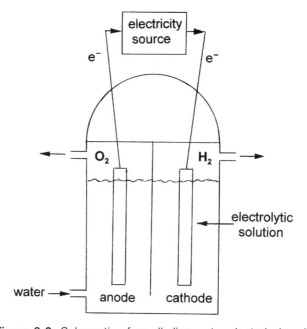

Figure 8-3. Schematic of an alkaline water electrolysis cell.

would be the use of an inorganic membrane in the cell that would increase the efficiency somewhat more (to about 90%) and reduce the electricity requirement to about 42 kWh/kg.

Solid Polymer Electrolysis   The use of electrolysers with a solid polymer electrolyte (SPE) makes an important change in the physical equipment for the production of hydrogen from water. This involves replacing the corrosive KOH electrolyte with a solid membrane between the electrodes that is porous to ionized hydrogen atoms (protons) and impervious to their electrons. Several membrane substances have been investigated for potential application in fuel-cell technology that also would be useful in electrolysers. They include the early patented Nafion (perfluorosulfonic acid), carbon fiber paper, and proprietary materials that are used as proton exchange membranes (PEMs). The electrode reactions are

$$H_2O \Rightarrow 2H^+ + {}^1\!/_2 O_2 + 2e^- \qquad \text{at the cathode}$$

and with the protons that diffuse through the membrane,

$$2H^+ + 2e^- \Rightarrow H_2 \qquad \text{at the anode}$$

Solid polymer electrolysers are not yet in commercial use because of the high cost of the membranes and the high loading of platinum as catalyst.

High-Temperature Steam Electrolysis   A second type of solid electrolyte is constructed from ceramic substances, such as yttrium and zirconium oxides (Y_2O_3, ZrO_2), that can operate at temperatures greater than 1000°C in systems that can electrolyze water vapor instead of liquid water. This method has been termed the Hot Elly process [3], as described for the use of low-temperature steam (~200°C) from geothermal reservoirs. The advantage of using high-temperature steam lies in the thermodynamic relationship between enthalpy and temperature (used in Section 8.21a) in the dissociation of water by electrolysis:

$$\Delta H = \Delta G + T\Delta S \qquad\qquad (8.21)$$

where ΔH = change in enthalpy (total energy demand)
  ΔG = change in Gibb's free energy (minimum work needed to drive the reaction)
  T = absolute temperature (°K)
  ΔS = change in entropy (a thermodynamic property of the system, J/°K).

The total energy needed to electrolyze water, ΔH, has two energy sources: ΔG, the energy supplied by the electricity, and $T\Delta S$, the thermal energy. Figure 8-4 shows the relationship of the two components as a function of temperature. The data show the decrease in electric energy demand as the thermal energy input to the system increases. The shaded area between 800 and 1000°C is considered the Hot Elly temperature of optimum efficiency.

Thermal Methods   Thermal methods of hydrogen production may become a major way to reduce the cost of hydrogen so that it is competitive with fossil fuels. The most direct way would be the thermal decomposition of water by heating it. Unfortunately, large-scale dissociation of water requires temperatures in excess of 2000°C, which makes this process uneconomical with today's technology. However, hydrogen can be produced in several thermochemical cycles at temperatures below 1000°C that are attainable in modern electric power

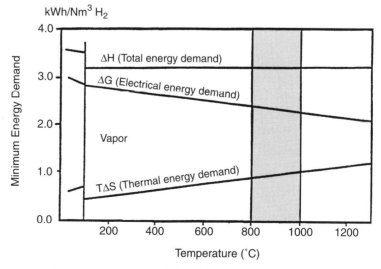

Figure 8-4. Minimum energy demand for high-temperature electrolysis of water [3].

plants. A thermochemical cycle utilizes a series of chemical reactions that liberate hydrogen while the chemicals involved are recycled in a closed system. A review of some 25 suggested thermochemical cycles was done in Brown and associates [4] from which two cycles were selected as most promising. The first was the sulfur-iodine (S-I) cycle (described by Besenbruch [5] and Schultz [6]), and the second was the Adiabatic UT-3 cycle (described by Yoshida and coworkers [7]).

The sulfur-iodine cycle is based on the high-temperature (800 to 1000°C) decomposition of sulfuric acid by the reaction

$$H_2SO_4 \Rightarrow \frac{1}{2}O_2 + SO_2 + H_2O \tag{8.22}$$

In the next step, the sulfur dioxide reacts with iodine and water by the exothermic reaction at a temperature of 120°C to form hydrogen iodide and regenerate the sulfuric acid by the reaction

$$SO_2 + I_2 + 2H_2O \Rightarrow 2HI + H_2SO_4 \tag{8.23}$$

In the second recycle step, the hydrogen iodide dissociates at a temperature of 320°C to produce molecular hydrogen and regenerate the iodine by the reaction

$$2HI \Rightarrow H_2 + I_2 \tag{8.24}$$

A schematic of the S-I process (with reaction temperatures) is shown in Figure 8-5.

The UT-3 cycle (developed at the University of Tokyo in the 1970s) is a four-reaction cycle based on the moderate temperature (300 to 750°C) reactions of hot gaseous bromine [$Br_2(g)$] and steam [$H_2O(v)$] flowing through fixed beds of calcium oxide and iron oxide in periodic reverse-flow processes with a net result of dissociating water into hydrogen and oxygen. The cycle starts with the liberation of oxygen from the calcium oxide bed with gaseous bromine at a temperature of 600°C by the reaction

$$Br_2 \text{ (g)} + CaO \text{ (s)} \Rightarrow CaBr_2 \text{ (s)} + \frac{1}{2}O_2 \text{ (g)} \tag{8.25}$$

The $CaBr_2$ remains on bed as a solid (s) that "splits" water into separate compounds, restoring the CaO at a temperature of 750°C and forming HBr by the reaction

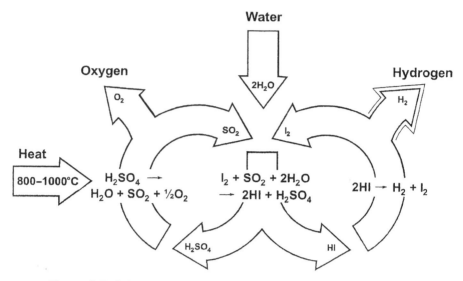

Figure 8-5. Schematic of the sulfur-iodine thermochemical cycle [8].

$$CaBr_2 \text{ (s)} + H_2O \text{ (g)} \Rightarrow CaO \text{ (s)} + 2 \text{ HBr (g)} \qquad (8.26)$$

In the separate fixed bed of iron oxide, the bromine is regenerated with the formation of ferrous bromide on the bed at a temperature of 300°C by the reaction

$$Fe_3O_4 \text{ (s)} + 8HBr \text{ (g)} \Rightarrow 3FeBr_2 \text{ (s)} + 4H_2O \text{ (g)} + Br_2 \text{ (g)} \quad (8.27)$$

On reverse flow, the hydrogen gas is liberated and restores the iron oxide at a temperature of 600°C by the reaction

$$3FeBr_2 \text{ (s)} + 4H_2O \text{ (g)} \Rightarrow Fe_3O_4 \text{ (s)} + 6HBr \text{ (g)} + H_2 \text{ (g)} \quad (8.28)$$

These cycles are under study to match their characteristics with optimum nuclear reactor types. For example, the type selected for the S-I cycle was the modular helium-cooled reactor described by Schultz [6] and the advanced high-temperature reactor described by Forsberg [8]. The more moderate minimum temperature required for the UT-3 cycle could be achieved with almost any type of nuclear reactor.

Biological Conversion Processes   Biomass conversion for hydrogen production has become an important aspect of the quest for renewable energy resources. One of the major arguments for raising public interest in replacing fossil fuels with hydrogen stemmed from the hope of stimulating public interest in coupling hydrogen production with renewable (green) energy resources. This was the theme of early books on hydrogen energy (e.g., [9]). It is convenient to classify renewable resources for hydrogen production into three groups: (1) solar energy for direct electric power, (2) solar energy by indirect processes, and (3) terrestrial resources for direct electric power.

Solar Energy for Direct Electric Power   Solar energy methods for direct electric power to produce hydrogen by electrolysis of water include hydroelectric dams, wind farms, and photovoltaic cells. These energy means were reviewed in Chapter 7, and the electric power output of any of these methods can be added directly to the electricity grid. Early consideration of photovoltaic systems (e.g., [10]) was for hydrogen production by electrolysis with an output of direct current (DC) electricity, as shown in Figure 8-6.

Two other concepts for solar energy production of hydrogen have been noted. The first is the direct production through photoelectrocatalysis in water splitting [11], in which an aggregation of small metallic

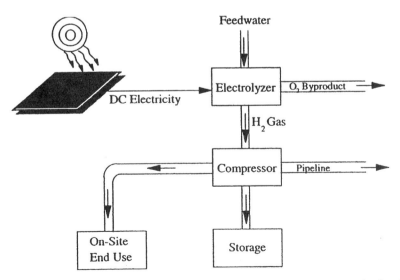

Figure 8-6. Schematic of a solar photovoltaic system for hydrogen production [10].

particles on less than 20% of the surface of the semiconductors allows about a 20-fold increase in the efficiency of the conversion of light to hydrogen. This concept is illustrated in Figure 8-7.

The second concept is a combined solar-hydrogen-battery electric grid system for the generation of electricity, its use, and storage control for chemically storing electricity in batteries or producing hydrogen as a fuel. This concept is illustrated in Figure 8-8.

Solar Energy by Indirect Processes   The alternative way to use solar energy to produce hydrogen is by biomass conversion to biofuels. The development of biofuels was covered in Section 7.4. An important advantage of using biomass for hydrogen production is the large hydrogen component (mainly as $C_nH_{2n}O_n$) of the biomass chemical composition, which allows direct conversion without an electricity generation step in the process. For example, in a biogas plant, the raw biogas can be processed into compressed methane for direct use as a transportation fuel. As shown in the sugar platform in Figure 7-15, ethanol can be produced for direct use as an additive to conventional fuels as well as being used for its thermal value to produce electric power and as a feedstock for hydrogen production by steam-ethanol reforming.

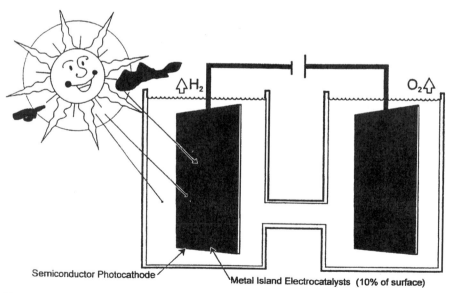

Figure 8-7. Schematic of a photoelectrocatalysis system for direct water splitting [11].

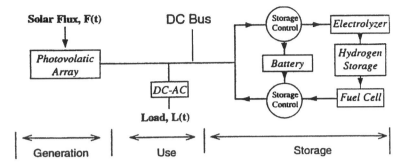

Figure 8-8. Schematic of a solar-hydrogen-battery electric grid system [12].

Another advantage of biofuels is their ease of use in small installations as a distributed energy source. Biomass as a feedstock in hydrogen production facilities could, in practice, make hydrogen a quasirenewable energy resource as well as an energy carrier. Some of the problems cited in its commercial development include the cost and sustainability of the biomass resource, the cost and suitability of the processing facility in the region, and the final value of the biofuels compared to the purchase and delivery cost of importing equivalent energy fuels from elsewhere. If steam-methane reforming becomes a major interim means of providing hydrogen fuel, biofuels could be a welcome addition.

Terrestrial Resources   Terrestrial resources suitable for producing hydrogen are geothermal steam deposits and uranium and thorium ores for the extraction of fissionable fuels for nuclear power reactors. Geothermal resources are widely distributed as geothermal steam deposits, making them a component of the energy mix in many nations. Geothermal energy, which makes a significant contribution to the U.S. energy supply, also can play a role in the production of hydrogen [13], as noted in a study of the potential for hydrogen production in the geothermal fields in Mexico. Several synergies have been listed, including (1) a more efficient use of the resources by providing an optimum balance between conversion to electricity for local grid use and hydrogen production for pipeline distribution as a fuel, (2) distribution of geothermal field investment over larger production, lowering the unit generation cost, (3) stimulation of geothermal resource development, and (4) increased reservoir lifetime by avoiding shut-in of geothermal wells to respond to cyclic demand as well as the more traditional reasons for reduction in the use of fossil fuels. The current slow development of geothermal steam deposits worldwide limits this resource to being a long-term option for the future of alternative energy sources.

The key terrestrial resource for hydrogen production is the fissionable isotopes of uranium and thorium distributed in mineral ores in the earth's crust. A discussion of the technology, resources, and reserves of these elements was given in Chapter 6 for both the open cycle (once-through use of the fuel rods, which then are discarded as waste) and the closed cycle (reprocessing of the essentially unspent expensively made fuel in the fuel rods, the produced radioisotopes that are useful as medical and industrial high-specific-energy sources, and the produced fissionable transuranic element by-products).

The human quest for abundant energy will make these high-specific-energy resources meaningful in the future, when public need overcomes public fear. One (or two) of the two possible results will occur in the future: (1) reextraction of the unreprocessed fuel from nuclear waste storage (or disposal) or (2) the development of civil applications of thermonuclear energy (solar energy on earth) that will complement (or displace) nuclear fission energy.

8.22 Hydrogen Storage

Electricity is the kinetic energy of moving electrons and cannot be stored as such. It can, however, be changed into potential energy in the form of chemical compounds (as in a battery) that release the electrons (on demand) at the battery voltage into a closed circuit until the stored chemical energy is spent. Hydrogen, unlike electricity, is a material form of potential energy that can be stored until needed. However at standard temperature and pressure (STP), the very low density of hydrogen (0.0899 kg/m^3 at STP) prevents its large-scale use except (perhaps) for stationary power generation. The corresponding specific volume of hydrogen at STP is 11.1 m^3/kg. Since a kilogram of hydrogen is approximately equal to a gallon of gasoline in automotive energy, a 15-gallon automobile tank would require a hydrogen gas tank at STP of 167 gallons of equivalent volume; this is clearly impractical for automobiles. There are three basic methods of hydrogen storage: compressed hydrogen gas (CHG), liquid hydrogen (LH_2), and chemical bonding, such as metal hydrides (MH).

Compressed Hydrogen Gas   Compressed hydrogen gas (CHG) is a commercial product; more than 5 BCM is used annually in the United States, mainly for synthesis of ammonia and hydrogenation of sulfur in petroleum refining. The gas commonly is sold in 50-liter steel cylinders pressurized to about 200 bar (~2900 psi). This pressure would be unsatisfactory for use in hydrogen fuel-cell automobiles. Four such cylinders, weighing about 270 kg (~600 pounds), would be needed to provide 3 kg of hydrogen, which would propel a small automobile about 400 km (250 miles).

The challenge for gas cylinder manufacturers is to develop light-weight materials for gas tanks holding about 5 kg hydrogen that would safely hold pressures up to 10000 psi (~680 bar), yielding a driving range of 400 miles at mean fuel consumption of 80 mi/kg. New light-weight composite cylinders have been developed [14] that can with-stand pressures up to 800 bar. Figure 8-9 shows the volumetric density of hydrogen inside the cylinder. The density of CHG under compression departs markedly from that of an ideal gas.

Electrolytic hydrogen is produced at pressures above atmospheric (P_0), and the theoretical work for isothermal compression of hydrogen is given by

$$\Delta G = R \; T \; \ln(P/P_0) \qquad (8.29)$$

where R = the gas constant (8.314 J/°K-mol)
       T = absolute temperature (°K)
       P = final pressure in the cylinder (bar)
       P_0 = initial pressure after production (bar)

Isothermal compression of hydrogen from 1 to 800 bar consumes 2.21 kWh/kg, but the actual energy consumed is larger because com-

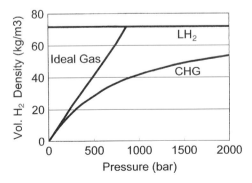

Figure 8-9. Volumetric density of compressed hydrogen gas as a function of gas pressure, including that of an ideal gas and liquid hydrogen. Source: Adapted from [14].

pression is not isothermal [14]. This adds a small ($<5\%$) fraction to the electrolysis energy consumption of $\sim$50 kWh/kg for producing the hydrogen fuel.

Liquid Hydrogen Liquid hydrogen (LH_2) is also a commercial product; it is used as a rocket fuel with liquid oxygen (LOx) in the space program. As shown in Figure 8-9, the density of liquid hydrogen is 70.8 kg/m^3 (at a temperature of 20°K).

Liquid hydrogen generally is produced in a multistage process with liquid nitrogen and a sequence of compressors and is stored in cryogenic tanks at ambient pressure. Liquid hydrogen can be stored only in open systems because of *boil-off*. Storage is complicated by two physical processes that result in evaporation (with a large increase in pressure to more than 100 bar) in a closed system. One is the heat leaks resulting from the size, shape, and thermal insulation of the storage tank. The boil-off loss from heat leaks is proportional to the surface-to-volume ratio; the evaporation rate diminishes as the size of the storage tank increases. For double-walled, vacuum-insulated spherical tanks [14], daily boil-off losses are typically 0.4% for a storage volume of 50 m^3, 0.2% for 100 m^3, and 0.06% for 20000 m^3. The second boil-off process results from the ratio of ortho- to para-hydrogen (see Section 8.01). Ortho-hydrogen in a storage tank at 20°K would slowly and spontaneously convert to para-hydrogen over a period of days to weeks, releasing enough transitional heat to evaporate much of the liquid hydrogen.

Although the theoretical energy required to liquefy hydrogen is about 3.2 kWh/kg, the actual energy requirement is about 13 to 14 kWh/kg; thus, the energy cost for storing electrolytically produced hydrogen is about 53 kWh/kg for CHG at 800 bar and about 64 kWh/kg for LH_2. At a cost of electricity of, say, 5¢/kWh, the electricity cost of hydrogen fuel would be about \$2.65/kg for CHG and about \$3.20/kg for LH_2. With the additional energy cost of makeup hydrogen to replace boil-off loses, it appears that the early applications of LH_2 will be in aviation and space, where cost is not the major criterion and consumption occurs in short periods. Still, the use of liquid hydrogen is being evaluated by several automobile manufacturers.

Chemical Bonding Chemical bonding is a way to store hydrogen in solid form in storage tanks in which the *filling* system places hydrogen atoms in the structure of the bonding chemical matrix at low temperature and the *operator* releases hydrogen gas as *fuel on demand* at elevated temperature. The more promising bonding chemicals are metals (such as magnesium, nickel, titanium, and their alloys) that can form covalent bonds with hydrogen to form metal hydrides by the reaction

$$2M + H_2 \Rightarrow 2MH + \Delta H(\text{heat of reaction}) \qquad (8.30)$$

The reaction is reversed at high temperature. Some of the metal (and alloy) hydrides under development for hydrogen storage, such as MgH_2, Mg_2NiH_4, $FeTiH_2$, and $LaNi_5H_6$, have higher volumetric density than does pure hydrogen. Although the weight percent of hydrogen in the metal hydrides is low (1 to 8%) compared with pure hydrogen (100%), the volumetric atomic density (atoms/m^3) is higher (5 to 7 compared with 1 for compressed hydrogen gas at 200 bar).

Other materials under study for bonded hydrogen storage include pure carbon in matrices such as activated carbon and large carbon molecules (e.g., buckminsterfullerenes, 60 carbon atoms in a ball structure called BuckyBalls). On the smaller side, carbon as nanotubes and glass microspheres are under investigation for hydrogen bonding.

8.23 Distribution of Hydrogen

The infrastructure for large-scale distribution of hydrogen fuel is not yet developed. The potential growth of a hydrogen fuel-cell vehicle industry in the world is still uncertain. Many automobile manufacturers have endorsed the concept of replacing the internal-combustion engine with fuel cells, and many oil companies have hydrogen fuel subsidiaries. However, the infrastructure for solving the "which comes first, the chicken *or* the egg" dilemma of when there are enough fuel-cell vehicles on the road to warrant widespread construction of hydrogen refueling stations and when there is enough distributed hydrogen fuel to warrant mass production of fuel-cell vehicles remains to be developed. Hopefully, the dilemma will be resolved together as a 'chicken *and* egg' solution. One attempt to resolve the dilemma is the "California Hydrogen Blueprint Plan" [15] for a "Hydrogen Highway Network" with the objective of installing a network of 250 hydrogen refueling stations in California for 20,000 hydrogen-fueled vehicles by 2010 to help set the stage for full-scale commercialization. The fate of this plan can be followed at www.hydrogenhighway.ca.gov/.

There are a number of methods for providing on-site production of hydrogen fuel where needed, such as on-vehicle production, local production for centrally refueled vehicle fleets, and small-scale electrolyser units for homes, home garages, and buildings. For large central facility production, a regional or nationwide distribution system of conventional refueling stations selling gasoline, diesel, and hydrogen fuel will be used. For distributed facilities, delivery of hydrogen fuel can be made by pipeline and tanker ships, railcars, and trucks.

For on-vehicle production of hydrogen fuel, the abundant distribution of liquid natural gas (LNG), methanol (CH_3OH), and other lightweight organic liquids made the idea of installing a chemical reactor into each automotive vehicle and small fuel-cell power plant to produce its own fuel by the steam reforming process look attractive. This was deemed a good interim method for supplying hydrogen fuel for fuel-cell vehicles until large-scale hydrogen filling stations became widely available during the transition period for large-scale production of fuel-cell vehicles. The practicality of this idea has been studied by the automobile manufacturing industry, and the conclusion thus far is that it is an impractical technology.

On-site production of hydrogen fuel for large fuel-cell power plants and large fuel-cell vehicle fleets that can be refueled locally at a central refueling station would make the need for distribution networks unnecessary. The production facility would become a reformer plant or electrolyser installation large enough to supply fuel demand at an appropriate rate as the user facility grew.

Large-scale use of fuel-cell power plants and private automobiles will require a hydrogen fuel distribution system equivalent to the present gasoline refueling (and maintenance) station network. The infrastructure to achieve this will entail using or adding to the existing natural gas pipeline network and developing a fleet of high-pressure or cryogenic tanker trucks and ships. The infrastructure also will require revision of existing standards and regulations to ensure the safety and security of extended networks of pipelines and shipping routes.

8.24 End Uses for Hydrogen Fuel

In Section 8.12, hydrogen was noted to have many competing uses. The concept of competing implies that there is not enough supply to meet the total demand. This is not true (yet) for hydrogen; production of industrial hydrogen is adequate to meet current demand. However, the demand for hydrogen as a fuel has barely begun except for its use as a thermal fuel in high-temperature welding equipment and as a vehicle fuel in the few fuel-cell research cars and demonstration buses that are running around a few cities for test purposes. The competition for hydrogen is evaluated in Chapter 10 in conjunction with the problem of the sustainability of electricity supply over the next 50 years as the fuel-cell and other energy-intensive industries develop.

There will be many applications for fuel cells, from miniature toys to large-scale power systems. The two major applications of hydrogen fuel will be for electric power generation and as a transportation fuel. The basics of the fuel cell and its use in transportation are discussed in Chapter 9. There are several applications for hydrogen fuel in electric power plants, large and small. A catalog of the uses would include

Load leveling
Standby emergency power supply
Small-scale, remote, and distributed power supply
Home systems

The concept of load leveling entails the use of excess rated capacity in off-peak periods to produce and store hydrogen for use in high-peak demand periods. The advantages of load leveling include operation in *base-load* mode, which is at a constant power output. Such operation avoids the need to build costly standby high-peak-demand power plants that generally are fueled with natural gas. The disadvantages include the cost of the electric energy used during the off-peak periods to produce the hydrogen. This application in the form of *dual-purpose* power plants is discussed more fully in Chapter 10.

A standby emergency power supply is needed in high-risk power-failure emergency situations, such as hospitals and municipal emergency facilities. Such a power supply is available with standby diesel combustion turbine units. Fuel-cell systems are now available commercially with standby power of up to hundreds of kilowatts. They generally are fueled with natural gas by the steam-methane reforming process.

For small-scale, remote, and distributed power supply, the technology for local production of hydrogen fuel in fuel-cell electric power plants is also available commercially. It would be useful for small communities and remote operation stations without maintenance personnel. The system needs a significant reduction in total investment cost, operational maintenance, and fuel cost to make it affordable in small remote communities. A new technology is being developed to replace currently used rechargeable batteries with micro fuel cells in portable appliances and personal electronic devices such as cell phones, laptop computers, and miniature TV sets. The fuel would be liquid methanol with minute flow from small containers that could greatly exceed the operating times of the best available batteries.

The technology for home systems is on hand, and installations are starting for in-home garage refueling of automobiles with small-unit electrolyser equipment. Figure 8-10 shows a small prototype electrolysis refueling system that was in operation in 2001 [16]. Systems also are becoming available for powering an entire home [17], such as the refrigerator-size 7-kW proton-exchange membrane fuel cell (PEMFC) appliance shown in Figure 8-11. A sketch of what a home power plant system [17] could look like (as of 1999) for both electrical and thermal services is shown in Figure 8-12.

8.25 Cost Factors of Hydrogen Fuel

The commercial success of all the applications cited above depends on the initial (first production) and mature (mass-produced) cost of the hydrogen fuel. Although these costs will not be known until production starts, it is possible to model the cost of electrolytic hydrogen on the

Figure 8-10. A commercial small-scale electrolytic hydrogen refueling appliance [16]. Courtesy of Hydrogenics Corp., Mississauga, Ontario, Canada.

Figure 8-11. A residential fuel cell appliance [17]. Printed with permission from A. C. Lloyd.

basis of its relative value to gasoline fuel. For example, in a study of the potential cost of hydrogen from geothermal steam resources, Fioravanti [18] examined the production cost of hydrogen as a function of four parameters relative to an energy equivalent cost of $23/GJ based on literature estimations. The parameters were input cost of electricity, type of electrolyser technology, pipeline distance for delivery, and, for purposes of emphasizing the value of air pollution abatement, the potential value for credit of emissions reduction. The results of this study are shown in Figure 8-13.

It is clear that the major cost of electrolytic hydrogen is the input cost of electricity with current electrolyser technology. For geothermal-derived electricity, this would be about 6 ¢/kWh. The potential for high-temperature electrolysis could reduce the cost to about 4 ¢/kWh. Pipeline delivery cost depends on both the distance delivered and the scale of production, but large facilities do not add greatly to the delivered cost. A key inducement to the introduction of hydrogen fuel for transportation would be the health value of reduced emissions. With the California Air Resources Board [19] estimate of the value of the cost of air pollution regulations of nitrogen oxides (NOx) and reactive organic gases (ROG) used in the introduction of zero-emission vehicle

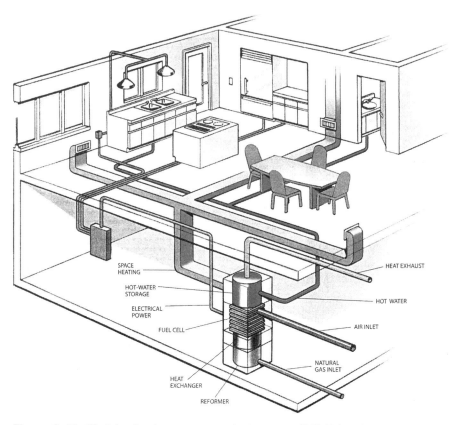

Labels within figure:
SPACE HEATING
HOT-WATER STORAGE
ELECTRICAL POWER
FUEL CELL
HEAT EXCHANGER
REFORMER
HEAT EXHAUST
HOT WATER
AIR INLET
NATURAL GAS INLET

Figure 8-12. Sketch of a home power plant system [17]. Printed with permission from A. C. Lloyd.

designation, the change in maximum electricity cost for delivered hydrogen at \$23/GJ would rise from just over 4 ¢/kWh to just over 6 ¢/kWh, making it more affordable.

8.3 SUMMARY

The chapter introduced the potential of hydrogen as a parallel energy carrier to electricity for both stationary power and transportation fuel applications. The fuel cycle aspects of production, storage, distribution, and end use were covered. The basics of electrochemistry were reviewed for large-scale electrolysis of water, the potential for reducing the cost of hydrogen by high-temperature electrolysis, and the possibility of thermochemical dissociation of water by several chemical cy-

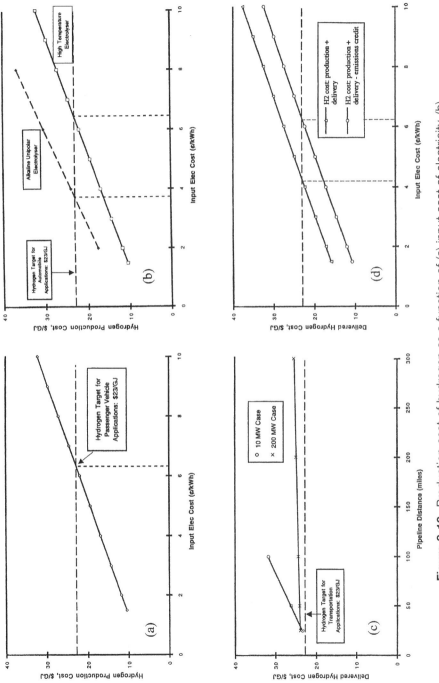

Figure 8-13. Production cost of hydrogen as a function of (a) input cost of electricity, (b) type of electrolyser technology, (c) pipeline distance for delivery, and (d) value of cost reduction of air pollution regulations [18].

197

cles. The aspect of storage was discussed in terms of the problem of hydrogen having a large specific combustion energy but a very small specific volume as a gas, thus requiring high-pressure gas tanks, liquefaction, or chemical bonding. The distribution aspect brought out the problem of local versus central production, and the end uses reviewed the concepts of large central power plants and residential units for home systems of automobile refueling and home energy systems. The chapter closed with a brief look at the cost factors most important in achieving an economical introduction of hydrogen as a transportation fuel.

REFERENCES

[1] D. S. Scott, "What Will We Gain?" *International Journal of Hydrogen Energy* 28:1031–37, 2003.

[2] S. Dutta, "Technology Assessment of Advanced Electrolytic Hydrogen Production." *International Journal of Hydrogen Energy* 15:379–385, 1990.

[3] V. K. Jonsson, R. Gunnarson, B. Amason, and T. J. Sigfusson, "The Feasibility of Using Geothermal Energy in Hydrogen Production." *Geothermics* 21:673–681, 1992.

[4] L. C. Brown, J. F. Funk, and S. K. Showalter, *Initial Screening of Thermochemical Water-Splitting Cycles for High-Efficiency Generation of Hydrogen Fuels Using Nuclear Power.* Report No. GA-A23373 of General Atomics Project No. 30047, San Diego, CA, April 2000.

[5] G. E. Besenbruch, "General Atomics Sulfur-Iodine Thermochemical Water-Splitting Process." *American Chemical Society,* Division of Petroleum Chemistry, Preprint No. 271:48–53, 1982.

[6] K. R. Schultz, "Use of Modular Helium Reactor for Hydrogen Production." Proceedings of the World Nuclear Association, Annual Meeting, London, September 2003.

[7] H. Yoshida et al., "A Simulation Study of the UT-3 Thermochemical Hydrogen Production Process." *International Journal of Hydrogen Energy* 15:171–178, 1990.

[8] C. W. Forsberg, "Hydrogen, Nuclear Energy, and the Advanced High-Temperature Reactor." *International Journal of Hydrogen Energy* 28:1073–1081, 2003.

[9] J. O. Bockris and T. N. Verziroglu, *Solar Hydrogen Energy: The Power to Save the Earth.* London: Macdonald & Co., 1991.

[10] J. M. Ogden and R. H. Williams, "Electrolytic Hydrogen from Thin-Film Solar Cells." *International Journal of Hydrogen Energy* 15:161–170, 1990.

[11] J. O. Bockris, "Hydrogen Economy in the Future." *International Journal of Hydrogen Energy* 24:1–15, 1999.

[12] S. R. Vosen and J. O. Keller, "Hybrid Energy Storage Systems for Stand-Alone Electric Power Systems." *International Journal of Hydrogen Energy* 24:1139–1156, 1999.

[13] M. D. Fioravanti and P. Kruger, "The Potential for Excess Geothermal Energy Capacity to Manufacture Hydrogen as a Transportation Fuel." Proceedings of the Tenth World Hydrogen Energy Conference, Cocoa Beach, FL, June 1994 (International Association for Hydrogen Energy, Coral Gables, FL, 1994), pp. 567–576.

[14] A. Zuttel, "Materials for Hydrogen Storage." *Materials Today,* 6:24–33, 2003.

[15] California Environmental Protection Agency, California Hydrogen Blueprint Plan: California Hydrogen Highway Network. Sacramento, CA: CEPA, May 2005.

[16] Hydrogenics Corporation, *Personal Fuel Appliance,* Stuart Energy Co., brochure, Mississagua, Ontario, Canada, 2002.

[17] A. C. Lloyd, "The Power Plant in Your Basement." *Scientific American* 281:80–86, 1999.

[18] M. D. Fioravanti, "An Analysis of Geothermal Hydrogen Energy Pathways." Goldman Interschool Honors Program in Environmental Science, Technology, and Policy, Stanford University: Standford, CA May 1994.

[19] California Air Resources Board, *Mobile Source Emission Reduction Credits,* concept paper, Scacramento CA: California Environmental Protection Agency, July 1992.

9

HYDROGEN AS A TRANSPORTATION FUEL

9.0 HISTORICAL PERSPECTIVE

Although hydrogen has been in use for more than a century, it has been used mainly as a chemical feedstock, not as an energy source. Hydrogen is employed for its thermal properties as a high-temperature fuel for welding. The main application of hydrogen as a transportation fuel was in the space program in the second half of the twentieth century. The purpose of its use as a space transportation fuel was to get the high-thrust propulsion needed to lift space vehicles out of the earth's gravitational field (which it did quite well). The combustion of liquid hydrogen (LH_2) with liquid oxygen (LOx) provided that thrust and kept the weight of the fuel low. Development of hydrogen production procedures (mainly the steam-methane reforming process) accelerated with NASA's need for large quantities of the two liquid fuels. The experience and lessons learned during that period initiated a wider interest in the potential of using hydrogen fuel for other forms of transportation: aviation, marine, and automotive. By the 1990s, national programs were under way in Germany [1], Japan [2], and, more slowly, the United States [3]. By the end of the century, the potential of fuel-cell engine vehicles with hydrogen fuel reached public awareness through the news media and presidential State of the Union (and state governor) addresses.

9.01 Hydrogen Fuel in Aviation

Liquid hydrogen fuel (LH_2) has long been considered an alternative turbine-engine aviation fuel [1,2]. Many technical requirements have to be met to establish the viability of LH_2 as a substitute for conventional jet fuel [4]. Two key factors for its use in aviation are (1) its contribution to aerodynamic efficiency, the ratio of lift to drag forces, and (2) its contribution to weight reduction of the fuel system (fuel and cryostats) for a meaningful reduction in fuel consumption. As a potential fuel for aviation, hydrogen must meet requirements for cost, availability, storage, volumetric energy density, safety, and ease in handling. The weight (and weight distribution) problem results from the need to use hydrogen fuel in the liquid state in cryogenic systems. Although hydrogen has high specific energy ($\sim$125 MJ/kg) (see Table 8-1), it has a very low volumetric density ($\sim$10 MJ/m^3 at STP). Even in the liquid state, hydrogen has a lower volumetric energy than gasoline (8.5:32 MJ/liter). Several studies are under way to examine the engineering factors for viability. At least two airplane manufacturers (Airbus and Boeing) have programs for the demonstration of airplane flight with hydrogen fuel-cell systems.

9.02 Hydrogen Fuel in Marine Technology

Although lagging behind the development of hydrogen fuel-cell applications for aviation and automobiles, the application of hydrogen for marine vessels has drawn interest. At least two projects have been announced for the investigation of hydrogen fuel-cell applications to provide electrical power and propulsion for marine vessels. One is the attempt in Iceland [5] to establish a national hydrogen fuel economy to replace that nation's dependence on imported oil. One of the key goals of that program is to switch to a fuel-cell fishing fleet, as fishing is important in Iceland's economy. The second is a high-technology consortium (the Maritime Hydrogen Technology Development Group) [6] that hopes to promote the use of hydrogen fuel cells to replace internal-combustion engines on watercraft of varying sizes. The first project studied by the group was the development and operation of a hydrogen-powered ferry to be operated by Pacific Marine in Hawaii. Another potential application was a hydrogen fuel-cell 18-passenger water taxi demonstration in 2003 on San Francisco Bay [7], powered by an Anuvu, Inc., fuel-cell/battery electric hybrid engine. Marine applications are also under study by the Office of Naval Research (ONR) [8] to develop propulsion systems that use fuel-cell technology for efficient generation of electrical power (and greater design flexibility) for future ships. ONR is funding the development of a method to extract hydrogen by diesel reforming. It is expected that the fuel-cell system will be capable of operating at 37 to 52% efficiency compared to the U.S. Navy's shipboard gas-turbine engines, which typically operate at 16 to 18% efficiency at the usual low to medium speeds that do not require peak use of the power plant.

9.1   HYDROGEN FUEL CELLS IN VEHICLE TRANSPORTATION

The worldwide interest in fuel-cell technology has focused on hydrogen as a transportation fuel in automotive vehicles. The major goal has been the replacement of fossil fuels in internal-combustion engines (ICE) with *clean-burning* hydrogen fuel in fuel cells. Italicizing "clean-burning" emphasizes the fact that the flame temperature of combustion of either gasoline or hydrogen in an ICE is $\sim 2400°C$ for the reaction

$$2H_2 + Air(N_2, O_2, Ar, ...) \Rightarrow 2H_2O + 2NO + ... \qquad (9.1)$$

This combustion temperature is well above the temperature of $\sim 1500°C$ for the reaction

$$N_2 + O_2 \Rightarrow 2NO \qquad (9.2)$$

whereas in a fuel cell, the electrochemical combination of hydrogen and oxygen takes place at a fuel cell temperature of $\sim 80°C$, which is too low for the production of nitrous oxides.

9.11   Just What Is a Fuel Cell?

The description of a fuel cell in countless publications is that "a fuel cell, in its simplest form, is a solid-state electrochemical system that combines hydrogen and oxygen to produce water and electricity." There are many excellent reviews, in print and on the Internet, of the basics of fuel cells. An example is the primer published by Los Alamos National Laboratory [9], as a good text for grades K–12. As a complement to the early full summary by Dutta (referenced in Section 8.21) of advanced methods of hydrogen production by electrolysis of water, the summary by Cannon [10] lists five types of fuel cells and describes many of their salient features, applications, and efficiencies as of about 1995. The five types are proton exchange membrane (PEM), phosphoric acid, alkaline, molten carbonate, and solid oxide fuel cells. The survey indicated that the PEM fuel cell is the type most promising for automotive transportation. Many reviews of the current status of fuel-cell technology are on the Internet, with more than 1.6 million hits on the inquiry "Fuel Cells."

The "still" photo of a single fuel cell, Figure 9-1, can be seen "working" at www.ballard.com. In place of the "light bulb" as the electric energy *load,* the overall efficiency of the cell is noted. In a fuel-cell vehicle (FCV), the fuel cell replaces the electric battery in an electric vehicle (EV) that provides the current for the electric motor that turns the wheels.

The section in Figure 9-1 labeled "Proton Exchange Membrane" in the PEM fuel cell allows the protons from the ionization of hydrogen at the porous anode

$$H_2 \Rightarrow 2H^+ + 2e^- \tag{9.3}$$

to diffuse through to the porous cathode to form water by the reaction

$$2H^+ + \tfrac{1}{2}O_2 + 2e^- \Rightarrow 2H_2O \tag{9.4}$$

An important problem in PEM fuel cells is the requirement for a platinum catalyst both at the anode for the ionization reaction and at the cathode for the recombination reaction to increase the reaction rate in both steps to achieve commercial validity.

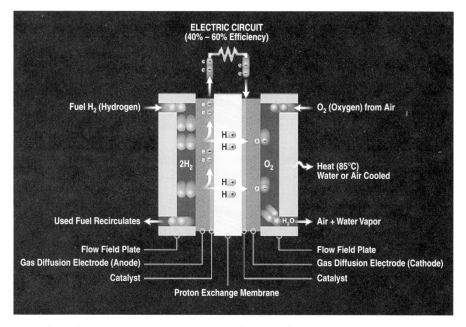

Figure 9-1. Schematic of a single working fuel cell. Source: Ballard Power Systems.

During the development period of fuel-cell vehicles, electric hybrid engines became available to reduce the consumption of gasoline (with the use of electric batteries) by means of a combination of onboard power generation and electric energy storage technology. Many ICE–electric battery hybrids (such as the Toyota Prius) are in production. Development is under way of fuel-cell–electric battery hybrids to achieve the same objective of increased fuel efficiency.

9.12 A Wee-Bit of Thermodynamics

This is a good place to examine the basic energy difference between the internal-combustion engine and the fuel cell. An internal-combustion engine is a *heat machine;* a fuel cell is not. The difference lies in two laws of thermodynamics that tell about the use of energy for doing *useful* work with energy resources.

Heat Engines   Heat engines are machines that convert thermal energy into kinetic energy, such as steam engines in locomotives and gasoline engines in automobiles. Heat engines follow the first law of thermodynamics, which states:

> "Energy may be transformed from one form to another but cannot be created or destroyed."

This means that the total energy of a closed system is constant. On the basis of this law, many people have sought the ideal *perpetual motion* machine that would run forever with no input of energy. Unfortunately, there is a second law of thermodynamics that states:

> "Energy transactions decrease the amount of useful energy available for further transactions."

This means that in addition to *quantities* of energy, there are *qualities* of energy. Energy quality ranges from *high-grade,* such as electricity, which is easy to convert to other forms, to lesser forms, such as *waste heat,* which is difficult to convert to *any* useful form.

The extent of transforms in the quality of energy is expressed by the *efficiency* of a thermodynamic process. Heat engines are limited in maximum theoretical efficiency, as deduced in 1824 by Sadi Carnot, who stated (approximately):

"The theoretical thermal efficiency (η) of the most efficient heat engine is fixed by the absolute temperature difference between the high-temperature and low-temperature heat sources in the engine."

This is expressed as the *Carnot cycle* theoretical maximum efficiency by the equation

$$\eta = (T_2 - T_1)/T_2 \qquad (9.5)$$

where T_2 = absolute temperature of the high-temperature heat source (K)

T_1 = absolute temperature of the low-temperature heat source (K)

For a fossil fuel power plant, converting chemical energy to thermal energy to produce steam by boiling water to mechanically drive a turbine-generator to generate electricity, the high-temperature source is the steam from the boiler [say, at a temperature of 547°C (820°K)] and the low-temperature source is the water from the steam condenser [say, at a temperature of 27°C (300°K)] after it passes through the turbine. For this power plant, the maximum Carnot cycle thermal efficiency would be

$$\eta = (800 - 300)/800 = 0.63 \ (63\%) \qquad (9.6)$$

In practice, the actual efficiency is much smaller because thermal inefficiencies in heat transfer losses from the system (boiler, steam pipes, valves, etc.) reduce the overall efficiency of the conversion of chemical energy to *useful* thermal energy to about 35 to 40%.

Fuel Cells   The fuel cell, as an electrochemical device, follows the same two laws of thermodynamics but gets around the limitation of the Carnot cycle by direct conversion of chemical to electric energy, eliminating the intermediate conversion of chemical energy to thermal energy. A comparison of the two conversion processes is shown in Table 9-1. The major losses in fuel-cell conversion efficiency are the losses in the reforming process, losses from the fuel cell operating at about 80°C, and losses in the AC/DC current conversions.

The theoretical efficiency of fuel cells is determined by the combustion energy of the fuel (the change in enthalpy of the fuel, ΔH) and

Table 9-1 Comparison of heat engines to fuel cells

Energy Conversion	Heat Engine Efficiency loss (%)	Fuel Cell Efficiency loss (%)
Chemical to thermal	60–70	—
Thermal to mechanical	1–2	—
Mechanical to electric	1–2	—
Chemical to electric	—	40–60

the energy that can be extracted as other than thermal energy (the change in entropy at the process temperature, $T\Delta S$) by Eq. 8.21. For the overall electrochemical reaction

$$H_2 \ (g) + \tfrac{1}{2}O_2 \ (g) \Rightarrow H_2O \ (l) \tag{9.7}$$

the enthalpy change ΔH is -285.8 kJ/mol (an exoergic release of energy) and the change in Gibb's free energy is $\Delta G = -237.1$ kJ/mol. The maximum theoretical efficiency (in percent) is

$$\eta = \frac{\Delta G}{\Delta H} \times 100 = 83\% \tag{9.8}$$

The voltage attained by the fuel cell is determined by the nonthermal energy Gibb's free energy for the half reaction at the anode

$$H_2 \Rightarrow 2H^+ + 2e^- \tag{9.9}$$

is given by the equation for the Gibb's free energy (in kJ/mol):

$$-\Delta G = n \ F \ E_o \tag{9.10}$$

where n = number of moles of hydrogen (= 2)
 F = Faraday constant = 96,485 coulombs/mol
 E_o = cell voltage (volts)

Thus, the ideal cell voltage is $E_o = -\Delta G/nF = (-237.1 \times 1000)/(2 \times 96485) = 1.23$ V.

For fuel-cell operation at 80°C (353°K), the change in the thermal component ($T \Delta S$) for the change in temperature from room temperature at 25°C (298°K) is 55°K, which reduces the maximum cell voltage to 1.18 V. Other effects in the cell, such as the use of air instead of oxygen and the humidity of the gases, reduce the maximum voltage still further to about 1.16 V. The actual operating voltage achieved in PEM fuel cells is about 0.7 volt, representing about 60% of the ideal maximum voltage.

9.13 Aspects of Hydrogen as a Transportation Fuel

In the selection of a transportation fuel, three sets of criteria come into play. The choices involve:

An optimum energy carrier from acceptable primary energy resources

Acceptable vehicle characteristics from the energy carrier

Infrastructure for fuel distribution.

Primary Energy Resources for Hydrogen   The primary energy resources for hydrogen production have been explored in earlier chapters. Cannon [10] prepared a summary of them, which is given in Figure 9-2.

It is clear that for the internal-combustion engine, the comfort and ease choice will remain gasoline (and diesel) as long as the supply is available and LNG if it is produced in large quantity. It was noted that

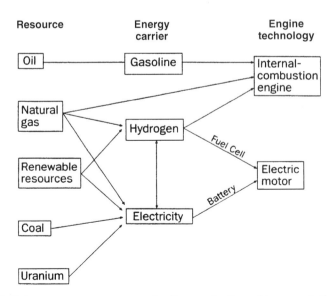

Figure 9-2. Primary energy resources for the production of transportation fuels [10].

other candidate fuels in the fossil and nuclear fuel categories include higher hydrocarbons such as methanol, ethanol, and propane and thermonuclear fuels (the heavier isotopes of hydrogen). The selection aspects of the choice include technical feasibility, safety, environmental impact, and, of course, economics.

The vehicle characteristics for the fuel involve the technical aspects of the

1. Weight of an onboard energy storage system (i.e., the fuel and fuel system hardware)
2. Propulsion efficiency (e.g., mean distance between fuelings)
3. Distribution of space (between equipment, passengers, and luggage)
4. Speed of refueling (measured against the current time for gasoline)

The other selection aspects are under study. One of the key ones already discussed is the choice of hydrogen storage and delivery. Although at this time the most likely choice is compressed hydrogen gas (CHG), it is not clear that research breakthroughs in solid metal hydrides or other solid (or liquid) bonding agents will make hydrogen storage more attractive and, hopefully, more economical. Safety considerations will always be at the forefront. The development of codes and standards is well under way on an international basis to avoid having conflicting national standards impede international trade.

Environmental impact analysis of the hydrogen fuel cycle (based on a *fair* comparative analysis of "from wells to wheels") is under way. Here the term *wells* refers to oil and natural gas resources. Although it generally is agreed that hydrogen fuel cells produce essentially only water vapor from the tailpipe, a *fair* fuel cycle analysis depends greatly on the choice of primary energy for producing hydrogen. Here the divisions quickly form between the Greens ("It's our chance to exist with renewable energy only"), the Petroleum Haves ("There's no foreseeable shortage of fossil fuels that will only become cleaner"), and the Nukes ("It's the only fuel that can satisfy the whole rapidly growing large-scale demand for electricity and transportation"). There will be more on this in Chapter 10.

9.14 Hydrogen Fuel Vehicles by Application Type

A problem in *getting going* on the final development of PEM fuel-cell engines is the large number of vehicle types now in production, from small scooters to large semis. Each type needs a specific power for optimum performance. Thus, fuel-cell engines must be designed for each type of vehicle that will be in production during the transition period. Another problem is the ease of introducing fuel cell vehicles by application usage as the infrastructure for hydrogen fuel distribution grows. The key to this problem is market development by type of refueling. For this purpose, it is convenient to consider motor vehicles in four weight-usage applications:

Automobiles: lightweight vehicles used for the convenience of individuals. Examples are private cars, taxis, motorcycles, and rental cars.

Vans: lightweight vehicles used for local delivery of people, equipment, and goods. Examples are airport limousines, ambulances, mail carriers, repair parts, and packages.

Buses: heavy-weight vehicles used for the transport of people. Examples are school buses, city and intercity transit buses, trailers, and mobile homes.

Trucks: heavy-weight vehicles used for heavy goods and materials. Examples are rigs, semis, cement mixers, moving vans, and road construction vehicles.

The chronological order in which hydrogen fuel-cell vehicles will come into mass production while the infrastructure for fuel delivery and distribution develops will rest largely on the early focus on central versus distributed refueling. It is clear that the earliest demand for large orders of vehicles will come for fleets used for local travel over fixed daily work periods with the ability to refuel overnight at convenient refueling stations. This has been part of the basis for the bus demonstration projects for mass transportation in some 15 cities in the world. Although individual electrolysis units for home garage refueling of passenger cars exist today, the constraint of local travel makes the early marketing of passenger cars less inviting.

9.2 HYDROGEN FUEL-CELL VEHICLES

Fuel-cell engines are manufactured as *stacks* of cells, with each cell adding to the total electric power of the engine designed to drive the wheels of the motor vehicle. The net voltage (E) generated by a single PEM fuel cell is ~0.7 volt, too small to propel an automobile. The operating voltage of a vehicle fuel cell stack is the number of cells in the stack times 0.7 volt/cell. The current density (I) at 0.7 volts is ~0.5 A/cm², and the power output (P = IE) is about 0.35 W/cm². Thus, to provide power of 50 kW to a fuel-cell engine, the area of polymer electrolyte membrane has to be about 140,000 cm² (14 m²).

Commercial viability of hydrogen fuel-cell vehicles depends on the production of fuel-cell engines that can compete successfully with internal-combustion engines. Viability also requires successful engineering of the total automotive mechanical assembly. The fuel-cell stack integrated into the fuel-cell engine, including the fuel delivery system, needs to fit into the space and weight limitation of the vehicle type. A cutaway diagram of a hydrogen fuel-cell vehicle (a Mercedes-Benz A-Class F-Cell) [11] with a 65-kW engine) is shown in Figure 9-3. The fuel-cell stacks and two 350-bar tanks for compressed hydrogen fuel are located below the seats.

A schematic of the total mechanical system of a hydrogen fuel-cell vehicle (the Ford P2000) [9] is shown in Figure 9-4. Here the fuel stack is next to the compressed hydrogen gas tank, and the electric motor drives the wheels.

A new concept for fuel-cell automobiles was announced by General Motors [12] as the "AUTOnomy Hydrogen Fuel Cell Vehicle" in which all of the propulsion system (hydrogen fuel, fuel cells, heat exchangers,

Figure 9-3. Schematic of the Mercedes-Benz A-Class F-Cell hydrogen fuel-cell vehicle. Source: Ballard Power Systems.

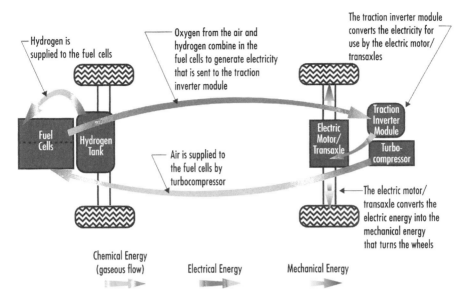

Figure 9-4. Schematic of the Ford P2000 hydrogen fuel-cell vehicle power system [9].

electronics, motors, suspension, steering, and brakes) are in a chassis (termed a "skateboard") as shown in Figure 9-5, to be sold with a choice of "tops" (such as sedan, sports car, SUV, etc.). The goal is the design and validation of a fuel-cell that is competitive with current ICE vehicles in durability, performance, and cost.

The more immediately feasible fuel-cell vehicle for widespread use is the bus. The bus has the previously mentioned advantage of local service with central refueling, the ability to hold a large volume of hydrogen fuel for daily consumption, and sufficient room for large fuel-cell stacks. A Ballard fuel cell bus used in many "on-the-street" demonstration bus projects is shown in Figure 9-6. About 40 advanced engineering versions of fuel-cell buses are under way in demonstration projects in more than 15 cities in North America, Europe, and Australia.

9.21   Characteristics of Alternative Fuels for Fuel Cells

Since the infrastructure for the production and distribution of hydrogen fuel may not be in place to match the potential growth of a fuel-cell industry, a transitional period of vehicle production with onboard re-

Figure 9-5. General Motors AUTOnomy hydrogen fuel-cell vehicle [12].

formers was considered. For onboard hydrogen production, the fuel system would include a mini-chemical plant component for reforming the primary fuel. Several candidate reformate (fossil) fuels were studied. A description was tabulated [9] of the characteristics of several of those fuels, which in addition to hydrogen included compressed natural gas (CNG), methanol (CH_3OH), ethanol (C_2H_5OH), and reformulated gasoline (RFG). The characteristics included aspects of production, storage, cost, safety, and environmental impacts. Because of the complexity of having a mini-chemical plant traveling aboard vehicles, the idea has been abandoned.

9.22 Methanol as a Fuel for Fuel Cells

Projects are under way at several national laboratories [e.g., Los Alamos National Laboratory (LANL) [9] and Jet Propulsion Laboratory (JPL)] to examine the feasibility of using methanol (easily handled as a liquid fuel compared with hydrogen) directly in fuel cells in place of hydrogen. These cells may be utilized as small (battery replacement)

Figure 9-6. A Ballard fuel cell engine bus used in "on-the-street" bus demonstrations.

mobile energy sources (1- to 10-watt applications) to greater than 1-kW stacks. A schematic of a direct methanol fuel cell (DMFC) is shown in Figure 9-7. Its use as automotive power plant in fuel-cell vehicles appears to lie well in the future.

9.23 Natural Gas as a Transportation Fuel

Compressed (CNG) and liquid (LNG) natural gas are currently in (more or less small) use as a transportation fuel in many countries. Its distribution as a household heating fuel and relatively high specific energy as a liquid allow it to compete strongly with hydrogen. From the comparison of specific and volumetric energy of natural gas and hydrogen (in Table 4-6), the advantage of volumetric energy clearly lies with natural gas (3.38 kWh/Nm^3 for CNG versus 0.64 kWh/Nm^3 for CH$_2$). However, it was noted in Section 4.4 that the reforming of natural gas

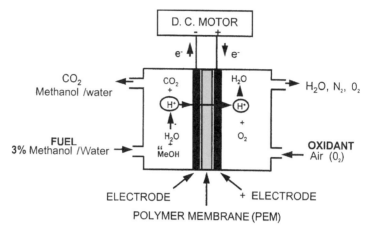

Figure 9-7. Schematic of a direct methanol fuel cell. Source: Jet Propulsion Laboratory.

to produce hydrogen by Eq. 4.3 for use in fuel-cell vehicles resulted in the use of 16 kg of methane to produce 8 kg of hydrogen. For their respective molar heats of combustion of 222 kWh of thermal energy for methane and 266 kWh for hydrogen, the question was raised: Why reform 16 kg of methane to produce 8 kg of hydrogen for a 20% increase in transportation energy instead of using the methane directly as a transportation fuel? This question will be reexamined in Chapter 10. It may be expected that during the ramp-up production of hydrogen fuel-cell vehicles over the next 50 years, if it becomes commercially viable, LNG and/or CNG may continue to compete strongly with gasoline in ICE vehicles.

9.3 WHAT MORE IS NEEDED?

The development of a hydrogen fuel-cell vehicle industry has been under way for more than 30 years. The advances in technology have been chronicled in many conference proceedings (e.g., the biannual meetings of the International Association of Hydrogen Energy [13] and the annual meetings of the National Hydrogen Association [14]), in journals focusing on hydrogen fuel technology (e.g., the *International Journal of Hydrogen Energy* [15]), in monthly newsletters (e.g., Hoffmann [16]), and on the Internet (the volume of information is growing rapidly). It was noted in Section 8.23 that the "chicken and egg dilemma" still exists on establishing the infrastructure of the industry. In addition, many aspects of the technology require improved (or new) methods, increased efficiency, and reduced cost. An early review of the technical issues was published by Cannon [10] in 1995. Since then, several road maps have been prepared, based on the current state of technology as described in the publications cited above to resolve the technical, environmental, economic, and safety problems for a large-scale fuel-cell vehicle industry. A summary of the technical issues as of 1995, as described in Cannon [10], is given in Table 9-2. It illustrates the worldwide effort undertaken since that time to achieve this industry.

9.4   SUMMARY

The chapter examined the use of hydrogen fuel for replacing conventional automobile engines with fuel-cell engines. The basics of thermodynamics were reviewed to allow a comparison of the efficiency of these two types of engine by application type. The chapter examined

Table 9-2   Outline of technical issues for hydrogen fuel-cell engines[a]

Challenge	Notes	
Reduction of weight/power	Current ICE:	2–4 lb/kW
	Early fuel cells:	15–30 lb/kW
	1995 technology:	~4 lb/kW
	Long-term goal:	1–2 lb/kW
Reduction of catalyst cost	Platinum for electrodes large part of FC cost	
	1995 cost	~$500 per engine
	Search for lightweight common metal catalysts	
Greater durability	FC (like rechargeable batteries) degrade with time	
	ICE lifetime:	≥5000 hours
	1995 FC lifetime:	~few thousands hours
Reduction of contamination	ICE:	residues burned or passed
	FC:	deactivated with CO/CO_2
Operating temperature	Optimal PEM T reached quickly and maintained	
Control of water	Water dilutes reactants	
	FC needs to be kept moist	
	Need for water content at optimum minimum	
Improved engine start-ups	Need to start at turn of key with short warm-up period	
Flexibility of fuel	Choice of H_2 + air versus H_2 + O_2	
	Choice of pure H_2 versus onboard reformed H_2	
Reduction of power cost	ICE:	~$50/kW
	1995 FC:	~$1500/kW

[a] Adapted from Cannon (1995) [10].

some of the aspects of hydrogen in fuel-cell vehicles, reviewed the design of some of the fuel-cell vehicles under commercial development, explored a few alternative fuels for fuel-cell engines, and raised the question of what more is needed to get the fuel-cell vehicle industry going.

REFERENCES

[1] German Industrial Consortium, *Solar Hydrogen: Energy Carrier for the Future,* brochure, 1992.

[2] New Energy and Industrial Technology Development Organization, *WE-NET: World Energy NETwork,* brochure. Tokyo: NEDO, 1993.

[3] D. Morgan and F. Sissine, *Hydrogen: Technology and Policy.* CRS Report for Congress No. 95-540 SPR, April 28, 1995.

[4] R. O. Price, "Liquid Hydrogen—An Alternative Aviation Fuel." *Aerospace Engineering,* 11:21–25, 1991.

[5] B. Arnason and T. I. Sigfusson, "Iceland—A Future Hydrogen Economy." *International Journal of Hydrogen Energy* 25:389–394, 2000.

[6] R. W. Foster, "An Investigation of the Integration of Hydrogen Technology into Maritime Applications." Proceedings of the 2000 Hydrogen Program Review, NREL Report CP-570-28890, 2000.

[7] PRNewswire, "First Hydrogen Fuel Cell Water Taxi on San Francisco Bay Powered by Anuvu." Sacramento, CA: PRNewswire, October 16, 2003.

[8] Office of Naval Research, *Hybrids on the High Seas: Fuel Cells for Future Ships.* Washington, DC: ONR Media, February 26, 2004.

[9] S. Thomas and M. Zalbowitz, *Fuel Cells—Green Power.* Report LA-UR-99-3231. Los Alamos, NM, 1999.

[10] J. S. Cannon, *Harnessing Hydrogen.* New York: INFORM, Inc., 1995.

[11] A. J. Appleby, "The Electrochemical Engine for Vehicles." *Scientific American* 281:74–79, 1999.

[12] *San Francisco Chronicle,* January 19, 2002.

[13] International Association for Hydrogen Energy, "Hydrogen Energy Progress." *Proceedings* of the World Hydrogen Energy Conferences, Vols. 1–14, biannually through 2004.

[14] National Hydrogen Association, "Annual U.S. Hydrogen Conference." *Proceedings*, Vols. 1–16, annually through 2005.

[15] T. Nejat Veziroglu, ed. in chief, *International Journal of Hydrogen Energy,* Vols. 1–30. Oxford, UK: Pergamon Press, through 2005.

[16] P. Hoffmann, ed., *The Hydrogen & Fuel Cell Letter*, Vols. 1–20, (1986–2005).

10

THE HYDROGEN
FUEL ERA

10.0 PERSPECTIVE ON AN ERA

It is difficult to pin down an approximate starting date for the hydrogen fuel era after the discovery of hydrogen in 1776, but it is easy to say that as an *era* it has hardly begun. With more than half a century of accelerating development of the technology, the future of the hydrogen fuel era is still uncertain with respect to two major concerns: (1) the potential for severe climate change as a result of the greenhouse effect from fossil fuel combustion and (2) the sustainability of the world's energy supply.

The second concern would include the production of hydrogen as a transportation fuel as well as for the generation of electric power. The demand for both of these energy vectors could be growing continuously on a large scale. Perhaps a more pressing environmental driving force than the potential greenhouse effect for a hydrogen era is the greater potential for air quality improvement in heavily populated municipal air basins. In this more immediate concern, the value of air pollution abatement in reducing the impact on human health (see Chapter 5) can be estimated.

The axioms for the quest for abundant energy (see Chapter 1) suggest that the fulfillment of the promise of a hydrogen era could play an important role in meeting the objectives of both of these environmental concerns. The problem of sustainability of the primary energy supply for a hydrogen fuel era as well as for existing (and other new) demands will become especially important during the next 50 years as the need for resolution of the fossil fuel problem becomes acute. In closing this description of the human quest for abundant energy, these two aspects of the potential for a hydrogen era require a deeper examination.

10.1 POTENTIAL FOR AIR QUALITY IMPROVEMENT

It was noted in Section 5.3 that despite the passage of the National Environmental Protection Act of 1969, the emission of air pollutants in automobile exhaust (from 1970 though 1998), although reduced significantly, still requires extensive effort in regulation to maintain that improvement, especially in large metropolitan areas such as Los Angeles, where the automobile is considered a necessity of life. This history has established a continuous need for regulation of automobile exhaust emission at both the national and state levels. The objective of the regulation generally is stated as "to reduce the emission of harmful exhaust chemicals from the operation of motor vehicles to below acceptable standards."

10.11 Emission Standards

Emission standards are set by federal [U.S. Environmental Protection Agency (EPA)] and state [e.g., Air Resources Board of the California Environmental Agency (CARB)] agencies. The magnitude of emissions is expressed as *emission factors* in units of grams of emission of specific pollutants per mile of travel (gpm) as measured in standard *driving cycles* under fixed test procedures. The emissions are calculated as the product of travel activity for a given type of vehicle and the emission factor for regulated tailpipe exhaust pollutants measured in laboratory dynamometer facilities, as shown in Figure 10-1.

The model used by the California Air Resources Board is the Motor Vehicle Emission Inventory (MVEI) Model [1] (revised through 2002 and available on the Internet). The model calculates the emission inventory of seven pollutants (HC, CO, NO_x, PM, Pb, SO_2, and CO_2) by performing an in-depth analysis of the data for 12 broad vehicle classes segregated by weight, usage, fuel, and vehicle age. The input data are the results of the emissions tests (with correction factors), model year travel data, and vehicle travel data for the fleet. The architecture of the model is shown in Figure 10-2.

Federal emission standards (under the Clean Air Act Amendments) are specified in the Federal Test Procedure (FTP) run on laboratory

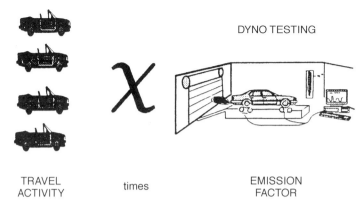

Figure 10-1. Calculation of pollutant emission under standard measurement conditions [1].

dynamometers for *standard* driving cycles. A driving cycle represents a typical history of an individual driver's acceleration of a vehicle from starting at rest to attaining *full* speed on a highway. Since each driver has an individual (and variable) driving cycle for reaching full speed (for congested city streets and for interstate highways), standard cycles are estimated for a number of full speeds. The amount of emissions released in a driving cycle is dependent on many factors. The problem in setting regulations is the range of significant discrepancies that exist between emissions measured under FTP conditions and *real-world* driving cycles.

In California, the annual on-road mobile source emissions inventory is calculated by the Air Resources Board with the EMFAC model [2]. The current version is EMFAC2002, which is detailed at www.arb.ca.gov. The model estimates the annual emission inventory resulting

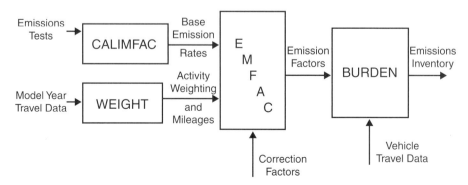

Figure 10-2. The Motor Vehicle Emission Inventory Model [1].

from 12 types of vehicles, the travel activity (as vehicle miles of travel (MVT) each year for each vehicle type), and the emission factors (adjusted for the ratio of city driving to highway driving). The model estimates the emission rates of 1965 and newer vehicles powered by gasoline, diesel, and electricity and can backcast and forecast inventories from 1970 to 2040. Results from the model are used for development of regulations, air quality plans to reduce pollution, and analysis of the impacts of transportation programs.

10.12   Factors That Affect Vehicle Emissions

As a major subject in environmental engineering, motor vehicle emissions have resulted in many research investigations. Six sources of in-use emissions have been established for accounting purposes in models and regulations:

1. Properly functioning warmed-up (hot-stabilized) cars in moderate on-cycle driving (on-cycle means driving described in the FTP)
2. Cold start for cars with properly functioning emission controls
3. Evaporation from the vehicle, including malfunctioning control
4. Off-cycle operation of vehicles with properly functioning emission controls (with a focus on higher power than is emphasized in the FTP)
5. Malfunctioning emission control systems affecting tailpipe emissions
6. Upstream emissions (from fuel extraction, transportation, refining, and distribution).

A review of real-world emissions by Ross and Wenzel [3] published in 1997 showed a complex network of lifetime emissions in the six categories with data for the model-year 1993 car and forecasts for 2000 and 2010 model-year cars. The lifetime emissions of three pollutants [CO, HC (hydrocarbons), and NO_x] of model-year 1993 cars were compared with the 1993 tailpipe standards. The forecast data for the model-year 2000 and 2010 cars were compared with the then (~1997) expected federal (EPA) and state (CARB) emission standards. The factors for lifetime emissions were weighted averages over vehicle life. The data for the 1993 cars showed that total emissions of CO and HC were four to five times the tailpipe standard and that emission of NO_x was about twice the federal standard and four to six times the California standard. The data for the 2000 and 2010 model-year cars showed that the expected emissions will meet the increasingly strict standards for on-cycle tests except for the California standards for LEV (low-emission vehicles) cars.

It is difficult to compare results from various studies over a period of years because the emission standards are in continuous evolution by the agencies that prepare them. It should be of interest to compare the 2000 forecast with the actual data for lifetime emissions from model-year 2000 cars, adjusted for changes in methods of measurement and emission standards. This is left for the reader to do as a deeper-look exercise.

10.13 History of California Emission Standards

The state of California, with the difficult task of meeting air quality standards mandated by the many amendments to the federal Clean Air Act, initiated a program in 1990 to intensify the path to improved air quality in that state by requiring all manufacturers of automobiles that wanted to sell new automobiles in California to meet a timed set of emission standards. The regulations promulgated in 1990 [4], known as the LEV Regulations, had three new primary elements in regard to emission standards: (1) tiers of exhaust emission standards for increasingly more stringent categories of low-emission vehicles, (2) a requirement for each manufacturer to phase in a progressively cleaner mix of vehicles each year, and (3) a requirement that a specified percentage of vehicles be zero-emission vehicles (ZEVs). The LEV categories later became the LEV-I categories, with definitions as follows:

TLEV	Transitional low-emission vehicle
LEV	Low-emission vehicle
ULEV	Ultra-low-emission vehicle
ZEV	Zero-emission vehicle

As originally adopted, the regulations required that specific percentages of the lighter vehicles be ZEVs by 1998. The LEV-I categories also carried a requirement for the standards to be met for a vehicle lifetime duration of 50,000 miles. Difficulties cited by the major automobile manufacturers in meeting those requirements led to revisions in the regulations [3], which now are known as the LEV-II standards. Major aspects of the revised regulations are summarized in Table 10-1. A category of emission-controlled vehicles was established for heavier-duty light cars (such as SUVs and pickup trucks), the super-ultra-low-emission vehicle (SULEV), for which the required durability

Table 10-1 LEV-II emission standards

Vehicle Category	Duration (miles)	NMOG (g/mi)	CO (g/mi)	NO_x (g/mi)
LEV	50,000	0.075	1.7	0.05
ULEV	50,000	0.040	1.7	0.05
SULEV	120,000	0.010	1.0	0.02
ZEV		0.000	0.0	0.00

was set at 120,000 miles. Also, emphasis was placed on three key pollutants: nonmethane organic gases (NMOG), nitrous oxides (NO_x) (both precursors of smog), and carbon monoxide (CO).

The first driving force to achieve zero-emission vehicles was the electric vehicle (EV). Demonstration projects by a few manufacturers showed that EVs were not practical, and they were taken off the market, to be replaced by hybrid gasoline-electric vehicles. The present nascent driving force is the hydrogen fuel-cell vehicle, which is in the development and demonstration phases. Further developments in emission standards can be found at www.arb.ca.gov/.

10.2 MODELING HEALTH BENEFIT FROM HYDROGEN FUEL TRANSPORTATION

Section 5.4 reviewed the major factors that determine the health benefit of reducing vehicle emissions of hazardous substances. It should be evident from the extensive scientific background needed to understand these factors that a lifetime of study by large numbers of scientists was (and still is) needed to make significant headway in calculating meaningful results for the ever-changing magnitude-severity conditions. Yet in Section 5.5 it was noted that examination of medical data and annual vehicle travel data in the Los Angeles metropolitan air basin resulted in an estimated epidemiological health value of 4.5 ¢/mile. Although the calculated value may not be significant to an accuracy of 0.1 ¢/ mile, it did help the municipal government realize the importance of reducing automobile exhaust emissions in the basin.

A major driving force in the encouragement of the use of hydrogen as a transportation fuel is its potential for eliminating vehicle emissions (other than water vapor). Although the value of reducing hazardous substance emission has only externality value, it was noted in Section 8.25 that the externality value is on the order of 2 ¢/kWh in the cost of electricity in regard to the economic viability of hydrogen fuel. This observation led to a model for estimating the health value of air pollution abatement resulting from the introduction of zero-emission vehicles in congested population centers, such as those shown in Section 5.3. Studies were initiated [5] for the metropolitan areas of Los Angeles and Mexico City and were carried out in detail for Tokyo under the WE-NET program in Japan.

10.21 Model Development for the Three-City Hydrogen Air Quality Study

The premise that quantification of epidemiological health benefit is difficult was based on the multiple types and sources of pollutants and the multiple pathways for public exposure. Quantification of the benefits is also difficult because of the problem of measuring small differences between large numbers, each with large uncertainties and each changing with time. Thus, the more specific problem of assessing the value of reducing emission of hazardous gases from automotive vehicles by replacing fossil fuels with hydrogen fuel could be solved with a more modest model based on an analysis of trends in population growth, vehicle ownership, travel, fuel economy, and emission factors.

Regulatory control of metropolitan automotive air pollution can be achieved by limitation of population, vehicle ownership, vehicle travel, and emissions. Since growth of population and vehicle ownership are socially difficult to control, air quality improvement requires a reduction in traffic activity and/or emission inventory. Limitation of traffic activity (e.g., by driveless Sundays and fuel ration coupons) is very unpopular (a symptom of the *do-without* philosophy). Thus, limitation of the emission inventory (e.g., by catalytic converters and exhaust inspections) is more publicly acceptable under Axiom 2. The replacement of internal-combustion engines with clean-burning hydrogen fuel in fuel cells (at least at the same cost and convenience) should be fully acceptable (a symptom of the *do-better* philosophy). The three-city study focused on governmental data of the four nonsocial parameters to estimate the trends in their growth and the potential for hydrogen fuel to assist in the do-better philosophy for achieving a significant reduction in automobile emissions.

The results of this study [5] showed several interesting differences in the trends for the three cities. The input data for the four parameters (obtained from the metropolitan governments of the three large urban areas) are reproduced in Tables 10-2 through 10-5. The three metro-

Table 10-2   Comparison of population trends

	Los Angeles	Mexico City	Tokyo
Metropolitan area (km^2)	17,100	4315	2183
Population (10^6)			
1970	9.61	8.68	11.41
1980	10.90	13.64	11.62
1990	13.86	15.05	11.86
2000 (estimated)	—	—	11.92
2010	17.40	20.00	—
Mean annual growth rate			
m.a.g.r. (%/a)	1.83	2.75	0.193
Doubling time (years)	37.9	25.2	365
Population density			
1990 (cap/km^2)	811	3488	5433
Extrapolations (at m.a.g.r.)			
2020	24.0	34.3	12.6
2050	41.5	78.4	13.3

politan areas are of meteorological interest, with Mexico City representing a closed air basin, Los Angeles a semiclosed air basin, and Tokyo an open air basin. There are large differences in area, population density, vehicle ownership, traffic volume, and air pollution problems among the three cities.

Table 10-3   Comparison of vehicle registration trends

	Los Angeles		Mexico City		Tokyo	
Metropolitan area (km^2)	17,100		4315		2183	
Vehicle registrations (10^6)	Autos	Total	Autos	Total	Autos	Total
1970			0.60	0.76		
1975					1.56	2.66
1980	6.04	8.04	1.78	2.10	1.81	3.09
1990	7.62	10.29	2.62	3.17	2.78	4.53
1994	7.59	10.15			2.96	4.58
2000 (estimated)			3.14	3.81		
Mean annual growth rate						
m.a.g.r. (%/a)	2.13	2.10	2.83	2.98	3.67	3.20
Doubling time (yr)	32.5	33.0	24.5	23.2	18.9	21.7
Vehicle density (cars/km^2)						
1990	446	602	606	735	1273	2075
Extrapolations (at m.a.g.r.)						
2020	14.4	19.3	5.15	6.24	8.40	11.8
2050	27.4	36.3	7.62	9.31	25.1	30.9

Table 10-4 Comparison of traffic activity trends

	Los Angeles			Mexico City			Tokyo		
	1990	2020	GR(%)	1989	2020	GR(%)	1990	2020	GR(%)
Vehicles (10^6)									
Automobiles	7.62	14.4	89.0	2.66			2.78	8.4	202
Trucks	2.61	4.9	87.7	0.31			1.75	3.4	94
Total	10.23	19.3	88.7	2.66		n/a	4.53	11.8	160
Traffic (10^9 VKT)									
Automobiles	121.2	214.1	76.7				32.85	40.60	23.6
Trucks	43.4	78.9	81.8				17.66	21.66	22.7
Total	164.6	293.0	78.0	n/a	n/a		50.51	62.26	23.2
Mean travel (10^3 km)									
Automobiles	15.9	14.9	−6.3				11.82	4.83	−59.1
Trucks	16.6	16.1	−3.0				10.09	6.37	−36.6
Fleet average	16.0	15.2	−5.0	n/a	n/a		11.15	5.28	−52.6

Values for 2020 are extrapolated at m.a.g.r.

Table 10-2 shows that Los Angeles has a smaller population density than do the other two although it does not have the smallest population. Tokyo, with the smallest metropolitan area, has the largest mean population density, and with a very small mean annual growth rate, the population density will remain essentially constant in the foreseeable future. Mexico City, in a closed valley air basin and with the largest population and the largest mean annual growth rate, will continue to increase in population density.

Table 10-5 Comparison of emission inventory trends (in kt/y)

	Los Angeles[a]			Mexico City[b]		Tokyo[c]		
	1990	2000	2010	1990	1994	1990	2000n	2000r
Hydrocarbons								
Automobiles	159.8	74.9	34.8	173.0	466.3	6.2	(s)	
Trucks	72.1	39.2	30.6	127.4	89.0	17.8		
Total	231.9	114.1	65.4	300.4	555.4	24.0	20.8	19.3
Nitrogen oxides								
Automobiles	105.7	61.2	39.5	51.5	60.0	17.5	18.3	17.9
Trucks	141.2	150.5	134.5	82.2	31.1	34.7	29.8	21.6
Total	246.9	181.7	174.0	133.7	91.8	52.2	48.1	39.5

[a] From CARB'96 [1].
[b] From DDF'95 [7].
[c] From BEP-TMG'94 [8]: n = with no additional restrictions; r = with double planned restrictions; s = small fractional vehicle contribution.

Table 10-3 shows the robust growth in motor vehicle ownership in all three areas over the 20-year period. The impact of growth in affluence [expressed in units of vehicles per capita (veh/cap)] of the three areas is noted in the data. The values for estimated growth from 1980 to 1990 (in veh/cap) were from 0.737 to 0.742 for Los Angeles, from 0.154 to 0.211 for Mexico City, and from 0.266 to 0.382 for Tokyo.

Table 10-4 shows many types of vehicles grouped into light (automobile) and heavy (truck) categories and their relative annual travel distances (in vehicle kilometers travel, VKT). The data for the Mexico metropolitan basin (involving three municipal governments) were not complete. The data show the Tokyo fleet, about half the fleet size of Los Angeles, increasing at about twice the growth rate. The mean annual travel per vehicle over the 10-year period decreased about 5% compared with the value for Tokyo, which decreased 10 times as much.

Table 10-5 shows the data for the two key precursors of photochemical smog: hydrocarbons and nitrous oxides. The 1990 data for Los Angeles were compiled from the California Air Resources Board Motor Vehicle Emissions Inventory (MVEI) Model [1], and the extrapolated data include assumptions of emission controls in the SCAQMD plan [6] of 1994. The data for Mexico City were obtained from two sources [7]; the connection between them was unclear. The data for Tokyo came from several Bureau of Environmental Protection–Tokyo Metropolitan Government reports (in Japanese) [8]. Forecasts for the year 2000 were for three cases of assumed regulations. The forecast reductions in the Los Angeles area were 50 to 72% for hydrocarbons and 24 to 30% for NO_x. The forecast reduction for hydrocarbons for Tokyo were not extrapolated because it was noted that the emissions inventory was small compared with the emissions from stationary sources. The reduction in NO_x of 8 to 20% was based on the proposed set of regulations. Because of the greater relative importance of NO_x over reactive organic gases (ROG), achievement of lower NO_x emission becomes more important.

In view of the more detailed data available for the Tokyo metropolitan area, the air quality abatement model for hydrogen fuel was applied to estimate the potential health benefit as part of the newly initiated WE-NET program [9] of the New Energy and Industrial Technology Development Organization (NEDO) of Japan.

10.22 The Metropolitan Tokyo Air Quality Study

The model described in Section 10.21 was applied to the severe air pollution problem in Tokyo by the Japanese WE-NET hydrogen development program [9] to examine the potential of hydrogen fuel to assist in the Tokyo Metropolitan Government (TMG) plans [8] for air quality improvement. The objectives of the study were to compile the database shown in Tables 10-2 through 10-5, to estimate the magnitude of future air pollution problems in the basin with respect to fuel type and influence of meteorological factors and to estimate the potential for emission reduction under various control regulations proposed by BEP-TMG [10].

The results of the study [11] were obtained with the Tokyo Metropolitan District Vehicle NO_x Model prepared specifically for the study as a graphical dynamic model (using the commercially available Stella II software described by Hannon and Ruth [12]). A schematic of the model is shown in Figure 10-3.

The model examined the emission of NO_x as a function of the vehicle fleet composition for the expected growth of gasoline- and diesel-fuel vehicles and the potential growth of hydrogen fuel vehicle (as zero-emission vehicles) production. The model used the emission factors and driving cycles provided by BEP-TMG to calculate the reduction in NO_x emission for several of the proposed regulations affecting travel activity in the Tokyo metropolitan area.

Figure 10-4 shows the potential growth in number of HZEVs (hydrogen zero-emission vehicles) as a function of mean annual growth rate (m.a.g.r.) from 20 to 50%/a. For the very ambitious goal of having immediate results by 2020, the end of the WE-NET planned study period, the curves show that a rate of 20%/a would have little effect, but at 50%/a the entire Tokyo vehicle fleet could be replaced with HZEVs.

Figure 10-5 shows the corresponding expectations for reduction in the emission of NO_x for one of the likely control plans. The curves reflect the expectations as a function of the m.a.g.r. of HZEV production. NO_x emission could have been eliminated in Tokyo by 2020 from a HZEV fleet introduced at a growth rate of 50%/a starting in 2005.

Figure 10-6 shows the results of the model for ambient NO_x concentration measured over the large network of monitoring stations in the Tokyo metropolitan area for the same proposed control plan. Any growth rate higher than about 30%/a would have been sufficient to reduce the roadside air concentration to below the 30 ppb (parts per billion) standard of the TMG, including about 30% of the ambient concentration of NO_x resulting from the nontransportation sector of the economy.

From available (incomplete) estimates of air pollution costs in Tokyo, the epidemiological health value for Tokyo would be in excess of

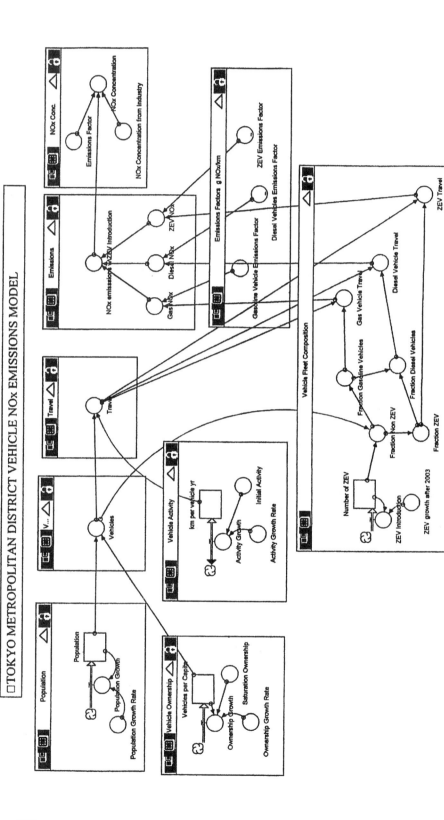

Figure 10-3. Schematic of the Tokyo metropolitan district vehicle NO$_x$ emission model [13].

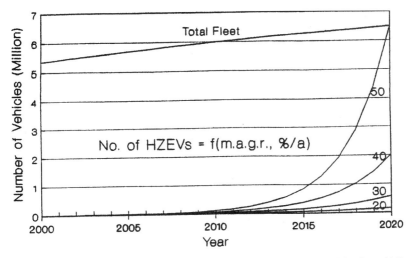

Figure 10-4. Potential growth of the HFleet in the Tokyo vehicle fleet [11].

0.4 ¥/km ($\sim$ 0.6 ¢/mi). The results of this study for the WE-NET program indicated a strong inducement to accelerate the commercialization of hydrogen fuel vehicle production (at least for use in the Tokyo metropolitan area).

The quest for a *clean and safe* environment with *comfort and ease* can be achieved at some cost, but since *avoidance* of adverse effects

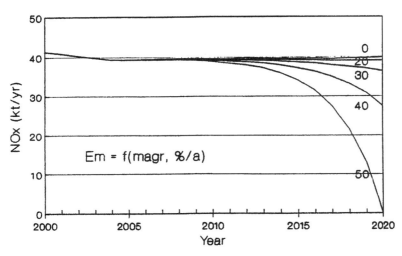

Figure 10-5. Potential reduction in NO_x emission as a function of HFleet growth rate [11].

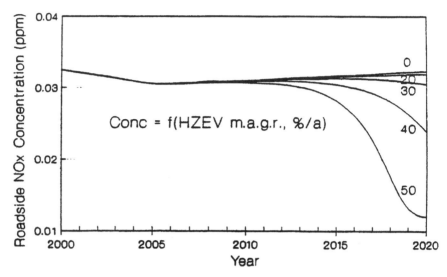

Figure 10-6. Forecast of NO_x concentration at roadside monitoring stations [10].

is not valued in today's marketplace (see Section 4.02), the externality value of avoiding the price paid for reduced public health may not lower that "some" cost enough to allow the transformation to a hydrogen economy. The need for cost reduction by technology improvement (i.e., abundant energy at low cost) is apparent, and the sustainable energy mix that would achieve that goal becomes the critical question.

10.3 ELECTRIC ENERGY REQUIREMENT FOR HYDROGEN FUEL

The conclusion of this examination of the human quest for abundant energy must respond to the question of whether a large enough energy supply will be available in the future not only to meet the growth in the business-as-usual worldwide economy but also for the development over the next 50 years of a new energy-intensive industry to replace fossil fuels in internal-combustion-engine vehicles with hydrogen fuel in zero-emission fuel-cell-engine vehicles. Although the two major driving forces of a desire for improved environmental quality and a sustainable energy supply support the objective of fuel replacement, the limitation of the supply of acceptable energy resources may make this goal unattainable. The total supply of industrial-scale energy must come from fossil resources, nuclear fuels, and renewable sources. An analysis of what this distribution might be in the next 50 years (with and without a hydrogen fuel industry) examines the possibilities.

Chapter 9 described the many developmental advances and demonstration projects of the fuel-cell vehicle future. The real initiation of a fuel-cell vehicle industry will be the dedication ceremony of the first large-output factory of fuel-cell vehicles under an infrastructure that provides sales and services, refueling stations, and all the other attributes of a large-scale industry. It is expected that such initiation might occur in the first decade of the new century. While the world watches the inevitable rise in oil and natural gas prices and the unceasing cry against greenhouse gas emissions, hesitancy continues by governments (and industry) to get projects under way. More research and development is *always* needed to convince people that investment is warranted. In the meantime, the longer the time to action, the longer the time to observable results. The *hestitancy* cost of relying on *comfortable* business-as-usual consumption of gasoline and diesel fuel can be estimated in a manner such as that shown in Section 4.3 for natural gas commitment for electricity generation, but on a much larger scale.

To examine the potential electricity requirement for producing hydrogen as a transportation fuel, the dynamic model for air quality improvement was refocused to investigate the fuel requirement for a hydrogen vehicle fleet. The result was a two-step dynamic model—the HFleet Electric Energy Demand Model—that examined the problem of energy sustainability. The first step in the model was used to extrapolate official (government) historical and forecast data to the year 2010, when it was expected that a fuel-cell industry would be under way and would expand rapidly. The second step was used to calculate, for a range of potential fuel-cell vehicle growth rates, the resulting hydrogen fuel requirement and the concomitant electrical energy and power requirements from 2010 to 2050, when a large fraction of a fleet could be operating with hydrogen fuel. Studies using this model subsequently were made for the state of California [14], the United States [15], New Zealand [16], and the world (revised) [17].

10.31 Extrapolation of Historical Transportation Fuel Data to 2010

Data obtained from government agencies and the Internet were compiled into time histories, and the mean annual growth rate (m.a.g.r.) for pertinent periods until 2000 was calculated by using regression analysis. The parameters for the first step of the model (and their respective units) are listed in Table 10-6. From these data, the parameter values for 2000 were set as the initial values for the model, and the m.a.g.r. was used for the extrapolation to 2010.

The model has been used to update the first estimates of the potential for a hydrogen fuel-cell fleet for the United States [15] and for the world [17], with more recent data through 2000 used to increase the precision of the calculated future electricity requirement for business-as-usual growth as well as for hydrogen fuel supply. Table 10-7 lists the data used for extrapolation to the initial year 2010 for the revised step 2 calculations of the scenario's extension to 2050 for the United States and the world. The travel values are in miles for the United States and in kilometers for the world for the given year. The values for fuel consumption for both are in U.S. gallons of gasoline as used by the U.S. Environmental Protection Agency in its annual listing of fuel economy for the world's motor vehicles.

Many of the vehicle data in Table 10-7 have greater uncertainties by vehicle type because of the reclassification of vans, such as pickups and SUVs, which were included under heavy vehicles earlier and under light (two-axle) vehicles in the revised studies. For the United States, the fleet values of ownership by size showed a 7% increase over the five-year change in data. The change in fuel consumption was 4% and the electricity values increased only 1% over the five-year-period difference. For the world data, the changes in the input data, based on the

Table 10-6   Cast of characters for step 1 of the model

Input Parameter	Units
Population	10^6 people
Population growth rate	%/a
Vehicle ownership (ith type of vehicle)	VpC(i)
Ownership growth rate	%/a (i)
Saturation ownership	VpC(i) (max)
Annual vehicle miles (or kilometers) traveled	VMT(i) (or VKT(i))
Vehicle fuel economy	mi (or km)/U.S.gal (i)
Electric energy load	PWh
Electricity demand growth rate	%/a
Installed power generating capacity	TW
Power capacity growth rate	%/a

Output Parameters	Units
Vehicle fleet (by type)	10^6 vehicles
Fleet fuel consumption	10^9 gallons (gasoline)
Electric energy requirement	PWh
Electric power capacity required	TW

Table 10-7 Input parameters for 2000 and extrapolated values for 2010

Parameter (Units)	United States			World		
	Initial Value (2000)	m.a.g.r. (%/a)	Output Value (2010)	Initial Value (2000)	m.a.g.r. (%/a)	Output Value (2010)
Population (10^9)	0.275	0.96	0.304	6.06	1.30	6.89
Ownership (VpC)						
Light vehicles	0.786	0.61	0.80	0.079	−0.62	0.078
Heavy vehicles	0.032	1.95	0.06	0.034	2.99	0.038
Fleet	0.818		0.86	0.113		0.116
Fleet size (10^6)						
Light vehicles	216		241	478	1.16	537
Heavy vehicles	9		18	206	2.48	264
Fleet	225		259	684		801
Travel distance	(10^{12} VMT)			(10^{12} VKT)		
Light vehicles	2.54	2.54	3.26	7.38	0.57	7.81
Heavy vehicles	0.21	3.53	0.30	1.78	1.30	2.03
Fleet	2.75		3.56	9.16		9.84
Fuel economy	(mi/gal)			(km/gal)		
Light vehicles	20.9	0.38	21.7	38.8	−0.34	37.5
Heavy vehicles	6.8	1.34	7.8	11.8	1.16	13.3
Fleet			20.5			32.5
Fuel consumption						
Gasoline (10^9 gal)			189			361
Electricity						
Energy load (PWh)	3.80	2.55	4.89	15.35	2.76	20.23
Installed power (TW)	0.81	1.50	0.94	3.40	2.09	4.19

periods 1980 to 1995 for the original study and 1990 to 2000 for the revised study, showed greater differences. Here again, many types of vehicles were grouped differently between light and heavy vehicles. The fleet size showed an 11% decrease over the five years. The difference carried over into the annual kilometers of travel and fuel economy for the heavy vehicles and thus indicated a large change in fuel consumption. The extrapolated electricity values showed a small change of a few percent over the five-year period, reflecting a slowly increasing demand for electricity.

10.32   Growth of Demand for Hydrogen Fuel and Electric Energy: 2010–2050

The more illustrative part of the model studies may be the results of the scenario analysis of the future for vehicle transportation with and without the replacement of fossil fuel with hydrogen fuel. The major choices in modeling a 40-year growth of a new industry are (1) when to start, (2) the initial production rates, and (3) the expected growth rates. The choice for (1) was fixed at 2010.

As was noted earlier, it was hoped that the driving forces of ensuring fossil fuel supply and resolving global warming concerns would overcome the barrier of hesitancy. The choice for (2) was selected as an initial production in 2010 of 10,000 each of light and heavy vehicles. The light-duty vehicles most likely would be centrally refueled fleets of automobiles and light commercial vans, and the heavy-duty vehicles most likely would be buses (and then trucks) shown from ongoing demonstration projects to be well suited for mass transportation (and heavy work) in congested urban cities. The revised model incorporated a logistic curve limit to the growth rate of the HFleet as it catches up asymptotically to the growing total fleet and with expected closure (or retooling) of ICE vehicle plants at the inflection point in time.

The choice for the key parameter (3), the growth rate of hydrogen fuel-cell vehicle production, was selected to cover the range of unimportant (20%/a), moderately important (30%/a), and fully successful (40%/a). The model incorporated steady technology improvement in electrolysis of water from the current mean consumption of about 50 kWh/kg of hydrogen gas in 2010 to a consumption of 40 kWh/kg by 2050.

Figure 10-7 shows the results of the model study in which the world vehicle fleet grows initially at the historical growth rate and reaches a size of 1.4 billion vehicles by 2050, continuing to grow at its then current mean annual growth rate.

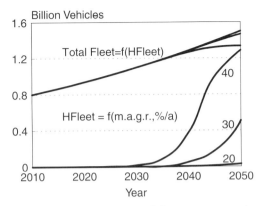

Figure 10-7. World HFleet as a function of the mean annual growth rate [17].

Figure 10-8 shows the corresponding hydrogen fuel requirement, which could be satisfied for the 20%/a growth rate by existing commercial hydrogen production facilities but would exceed 260 billion kilograms per year at total fleet replacement by 2050 at an m.a.g.r of 40%/a.

Figure 10-9 shows the resulting range of increased electric energy requirement relative to the business-as-usual growth in electric energy demand. The business-as-usual growth at 2.7%/a would be sufficient to include the energy requirement for the 20%/a growth rate, but replacement of the conventional fleet would require an additional 10 PWh/a by 2050. The additional installed capacity requirement would be about 1.35 TW, for a total installed capacity of about 10.9 TW.

A summary of the output data for both the U.S. fleet and the world fleet is given in Table 10-8. The growth of the hydrogen fuel fleet

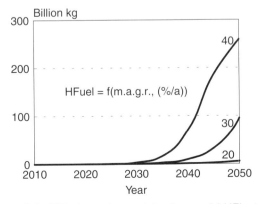

Figure 10-8. HFuel requirement for the world HFleet [17].

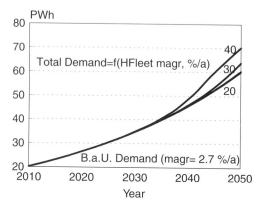

Figure 10-9. Total electric energy for business-as-usual growth and HFuel demand [17].

(HFleet) at the three scenario growth rates is compared to the business-as-usual growth of the total fleet. The marked difference in the hydrogen fuel (HFuel) requirement reflects the much smaller annual driving distances throughout the world compared with the United States but is offset in the later period by the larger mean annual growth rate in vehicle fleet size. These differences also are reflected in the corresponding requirement for additional electric energy and installed power capacity.

10.4 PROSPECTS FOR THE FUTURE OF A SUSTAINABLE ENERGY SUPPLY

The unending irreversible quest for abundant energy has been examined through this new century with reference to worldwide concerns such as the sustainability of fossil fuel supply, the effect of continuous release of greenhouse gases on world climate, the role of nuclear energy in the world energy mix, and the ability of renewable energy resources to provide a significant fraction of the future energy supply. The world's new concerns of how to deal with international terrorism using weapons of mass destruction and how to provide increased homeland security were not considered except to note the intensiveness of the energy needed to respond to those concerns. The studies in this chapter that estimated the energy requirement for moving from fossil fuel to hydrogen fuel in transportation raise the question of whether the world will move into a new era of parallel energy vectors of electricity and

Table 10-8 Summary of output data: 2010–2050

Year	M.a.g.r. (%/a)	Hydrogen Fleet		Hydrogen Fuel		Electric Energy		Power Capacity	
		United States (10⁶ veh)	World (10⁶ veh)	United States (10⁹ kg)	World (10⁹ kg)	United States (PWh)	World (PWh)	United States (TW)	World (TW)
2010	—	0.02	0.02	0.00	0.00	0.00	0.00	0.00	0.00
	B. as U.[a]	241	801	—	—	4.89	20.2	0.94	4.19
2030	20	0.77	0.77	0.22	0.14	0.01	0.01	0.00	0.00
	30	3.76	3.79	1.10	0.71	0.05	0.03	0.01	0.00
	40	16.1	16.5	4.71	3.09	0.21	0.14	0.03	0.02
	B. as U.[a]	318	1070	—	—	8.09	34.9	1.10	6.34
2050	20	27.6	28.9	8.09	5.39	0.32	0.21	0.04	0.03
	30	276	511	81.0	95.4	3.16	3.72	0.42	0.50
	40	357	1390	104	259	4.07	10.1	0.55	1.35
	B. as U.[a]	397	1500	—	—	13.4	60.1	1.45	9.58

[a]Forecast of business-as-usual vehicle fleet without hydrogen fuel-cell vehicles.

hydrogen, neither of which is a primary energy resource but both of which provide energy in an *ease and comfort* form. With the coming growth in population from 6 billion in 2000 to a forecast 9 billion in 2050, it appears that meaningful world planning will be needed in the very near future to ensure an adequate energy supply through the next generation or two under acceptable social and economic world conditions.

The data cited for the growth of world electric energy consumption indicate continuous growth in the need for additional electric power. The key uncertainties in evaluating the world electric power capacity that will be needed by 2050, with or without a hydrogen fuel industry, are (1) the actual growth rate of electric energy consumption and (2) the future role of nuclear energy in maintaining growth of the world economy. The forecast of world electric energy consumption by the U.S. Department of Energy [18] assumed a growth rate of 2.3%/a through 2025 compared with the historical growth rate of 2.8%/a over the last 10 years. The difference between these two growth rates represents a 28% increase above 41.0 PWh in 2025 (at 2.3%/a) to 52.4 PWh (at 2.8%/a) in 2050. This forecast also assumed that no new nuclear power plants would come on-line by 2025, which appears to hinder the needed long-term planning for an adequate electricity supply in the second quarter of the twenty-first century. In addition, the forecast through 2030 by the International Atomic Energy Agency [19] for electric energy and the proportional role of nuclear energy in the world shows a more uncertain picture. The forecast was a model of low and high estimates based on actual data for 2002 and a low estimate of a growth rate of 1.7%/a and a high estimate of 3.4%/a for total electric energy consumption. The growth rates for nuclear energy were 0.4%/a and 1.9%/a, respectively, also indicating a diminishing role for nuclear energy.

The question remains: Where will the additional energy required to change from fossil fuel transportation to hydrogen fuel as well as the business-as-usual growth in electric energy demand come from? The total world energy consumption must come from some combination of fossil fuels, renewable energy resources, and nuclear energy.

10.41 Potential Distribution of Energy Resources

Table 10-9 shows a potential distribution of energy resources that could meet the total energy requirement for 2050 with and without the additional electric energy needed for hydrogen production for a world

Table 10-9 Potential distribution of energy resources for world energy supply to 2050

Year	Forecast Demand IAEA[a] PWh	Model PWh[b]	Forecast Renewables[b] PWh	(%)	Fossil Fuels PWh	(%)	On-Line Nuclear[a] PWh	(%)
2002	16.1	16.2	2.89	(19)	9.9	(64)	2.57	(16)
2010	19.9	20.2	3.69	(18)	13.4	(66)	3.10	(15)
2020	27.9	26.6	4.62	(17)	17.9	(68)	4.01	(14)
2030	39.0	34.9	5.80	(17)	25.1	(72)	3.98	(10)
m.a.g.r. (%/a)	3.4	2.8	2.3	(−0.4)	2.2	(0.4)	1.6	(−1.2)
2050	n/a	70[c]	35	(50)	X		Y	

Range of Distributions for World Energy Supply with HFuel

X	(%)	+	Y	(%)	No.NPP
35.0	(50)		0.0	(0)	0
17.5	(25)		17.5	(25)	1750
0.0	(0)		35.0	(50)	3500

[a]Source: IAEA [19].
[b]Source: EIA [20].
[c]Total demand = 60 PWh (without H_2 production) + 10 PWh (with H_2).

HFleet. The distribution assumes that the world drive toward maximum use of renewable energy resources would result in a fraction of 50% of the total energy demand by the year 2050, with the other 50% obtained from a combination of fossil and nuclear fuels.

The table compares forecast demand through 2030 by the IAEA [19] high estimate case with the model results. It shows the rapidly growing deficit on the assumption of a 3.4%/a growth rate compared with the historical growth rate of 2.8%/a. The table shows the fraction of electric energy from renewable resources (mainly hydroelectric energy) declining from 19% to 15% at a rate of −0.4%/a and the negative growth in nuclear energy declining from 16% to 10% by 2030 at a rate of −1.2%/a. The table shows the model's estimate of total electric energy demand by 2050 reaching 70 PWh, consisting of 60 PWh from business-as-usual growth (including growth in the electronic age) and 10 PWh required to produce about 260 billion kilograms of hydrogen fuel per year for the fuel-cell vehicle fleet. It follows that if 50% of the total electricity demand could be met with renewable energy resources without large growth in additional hydroelectric power installation, the remaining 50% must be obtained from some combination of fossil-fuel and nuclear energy.

The second part of the table illustrates a range of distributions for these two energy resources, from all fossil fuel to all nuclear fuel. Long-term planning must consider issues such as the public's desire for a clean environment and renewable energy sources, the rapid expansion of a long-term commitment of natural gas combustion for additions to electric power capacity, the need to conserve natural gas for chemical feedstock and residential heating, and the additional electric power needed for the new created electricity demands, including a new hydrogen fuel industry that cannot become significant until after 2030, a period beyond current government planning.

The world now has more than 440 operating nuclear power plants with increasing efficiency and longer operating lifetimes. With the observation that a 1350-MWe nuclear reactor operating at a plant availability factor greater than 80% generates about 10 TWh per year, the approximate number of nuclear power plants (No.NPP) that could be needed ranges from 0 to about 3500, as reported by an environment correspondent [21] in *The Guardian.*

10.42   Possibilities to Resolve the Impasse

The study results show an acute need to plan for a larger than business-as-usual growth in electric energy demand over the next 50 years, especially if the hydrogen fuel age becomes significant after 2030, as shown in Figure 10-7. It would be difficult to abandon the production of fossil fuel vehicles over this 20-year time period, but substitution for fossil fuels by hydrogen fuel by either reforming or electrolysis must start if a 30 to 40 %/a growth rate of hydrogen fuel-cell vehicle production is to be achieved in the 40 years from 2010. The need for additional low-cost electric energy will be apparent in any case. There will be many pathways to achieve the transition to hydrogen fuel. Three ways to assist in the transition are suggested below.

The first is a change in the economics and design of new installed electric power plants. As suggested in 1994 [22] to the U.S. Department of Energy, the incremental cost of producing hydrogen by electrolysis of water at large power plants could be reduced significantly if the plants were designed for capacity well above the peak power demand instead of the normal few percent. The concept of a "dual-purpose power plant" is illustrated in Figure 10-10. With much of the operating and maintenance costs covered in the electricity cost, commercial quantities of excess electric power would always be available to produce hydrogen fuel at the marginal cost. The high-temperature cooling water

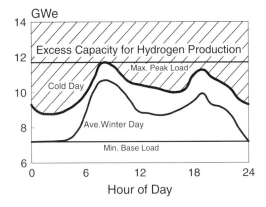

Figure 10-10. Daily demand curve for a dual-purpose electric power plant [22].

from such large plants could be used to preheat the electrolysis feed water, and this would synergistically reduce the electricity energy consumption per kilogram of hydrogen produced, as shown in Figure 8-4.

A second way to improve the availability of hydrogen fuel would be the realization of the potential to produce hydrogen by high-temperature thermochemical decomposition of water. The chemical basis of two chemical cycles for this technology was examined in Section 8.21, and the potential for large-scale commercial production of hydrogen fuel at high-temperature nuclear power plants might facilitate public acceptance of nuclear power as well as improve the hydrogen economy.

The third way to provide adequate electric power capacity would be the construction of "Solar-Nuclear-Hydrogen Energy Parks" [23] distributed among the many industrially developed and developing nations of the world. These facilities would be located in large industrial parks, located in remote high-solar-insolence areas, with a central cluster of nuclear power plants surrounded by a field of photovoltaic cells, possibly with a bank of wind power mills on the high side and the generator and electrolysis (or thermochemical) equipment near the delivery side. This synergistic coupling would reduce the problems associated with the choice of energy resources: the unpopularity of high specific energy from nuclear power (10^{11} kJ/kg) and the technical problems of low specific energy (10^4 kJ/kg) from solar power. These parks could be provided with "dual-purpose power plants," with a reduction in the cost of hydrogen fuel by means of solar preheating for higher-temperature electrolysis.

10.5 WRAP-UP

The human quest for abundant energy continues. Populations increase, technology develops, environmental concerns grow, and the economy grows to sustain all these changes. This will be true in the twenty-first century as it has been in the previous 20 centuries. History can tell us that humans, although anxious to build a better life of *comfort and ease,* show hesitancy to move into *uncharted waters.* But history also tells us that eventually they do move. The great technical strides achieved in the twentieth century, including the wartime introduction of a millionfold jump in specific energy, make for a high hesitancy barrier. It may take a generation or two to accept the risks of the future in regard to the complexity of the technical advances, but the axiom of irreversibility suggests that the large-scale potential of nuclear energy eventually will provide a major share of the energy mix of the future, especially when sustained "solar energy on earth" becomes a reality.

10.6 SUMMARY

The chapter examined in greater detail two key aspects of *getting going* with replacement of fossil fuels for thermal energy production. The first is the externality value of avoiding the risk of catastrophic climate change resulting from the greenhouse effect of fossil fuel combustion. The second, focused on completing the study of the human quest for abundant energy, is the problem of ensuring the sustainability of the world's energy supply as population and the quest for affluence continue to grow. The two aspects were examined with a home-grown dynamic model for the next 50 years. The results are startling, and three suggestions were given in the closing section on possibilities to resolve the impasse. The wrap-up to the book indicates that Axiom3 on *irreversibility* will dictate the need for twenty-first century energy sources of nuclear fission until the commercial development of "solar energy on earth" augments the commercial development of "solar energy from the sun".

REFERENCES

[1] Air Resources Board, *Methodology for Estimating Emissions from On-Road Motor Vehicles.* Sacramento: California Environmental Protection Agency, October 1996.

[2] Air Resources Board, *EMFAC2002: The Latest Update to the On-Road Emissions Inventory.* Sacramento: California Environmental Protection Agency, September 2002.

[3] M. Ross and T. Wenzel, "Real-World Emissions from Conventional Passenger Cars." Chapter 2 in J. DeCicco and M. Delucchi, eds., *Transportation, Energy, and Environment: How Far Can Technology Take Us?* Washington, DC: American Council for an Energy-Efficient Economy, 1997.

[4] California Air Resources Board, *LEVII and CAP2000 Amendments to the California Exhaust and Evaporative Emission Standards.* Sacramento: California Environmental Protection Agency, September 1999.

[5] P. Kruger, "Comparison of Potential for Air Quality Improvement from Hydrogen Fuel in Three Metropolitan Air Basins" *Proceedings* of the 11th World Hydrogen Energy Conference, Vol. 11. Stuttgart, Germany: International Association for Hydrogen Energy, June 1996.

[6] SCAQMD, *1994 Air Quality Management Plan.* Diamond Bar, CA: South Coast Air Management District, April 1994.

[7] DDF, *Bases de Politica Ambiental Urbana para el Manejo de la Cuenca Atmosferica del Valle de Mexico.* Comision Metropolitana para la Prevencion y Control de la Contaminacion Atmosferica del Valle de Mexico, Mexico City, October 1995.

[8] BEP-TMG, *Tokyo Metropolitan Government Plan to Prevent Automobile Pollution.* Tokyo: Bureau of Environmental Protection, July 1994.

[9] New Energy and Industrial Technology Development Organization, *WE-NET: World Energy NETwork,* brochure. Tokyo: NEDO, 1993.

[10] BEP-TMG, *TMG Automobile Pollution Control Plan.* Tokyo: Bureau of Environmental Protection, June 1997.

[11] P. Kruger, M. Murdock, T. Hirai, and K. Okano, "Potential for Air Quality Improvement in the Tokyo Metropolitan Area from Use of Hydrogen Fuel." *Proceedings* of the 12th World Hydrogen Energy Conference, Buenos Aires, Argentina, June 21–26, 1998, pp. 33–43.

[12] B. Hannon and M. Ruth, *Dynamic Modeling.* New York: Springer-Verlag, 1994.

[13] M. Murdock, *Predicting the Ultra-Low-Emissions Vehicle Introduction Necessary to Significantly Reduce NO_x Pollution Levels in Metropolitan Tokyo.* Graduate Research Project, Civil Engineering Department, Stanford University, Stanford, CA 1996.

[14] P. Kruger, "Electric Power Requirement in California for Large-Scale Production of Hydrogen Fuel." *International Journal of Hydrogen Energy,* 25:395–405, 2000.

[15] P. Kruger, "Electric Power Requirement in the United States for Large-Scale Production of Hydrogen Fuel." *International Journal of Hydrogen*

Energy, 25:1023–1033, 2000. Revised as "Potential Electric Power Capacity Required in the United States by 2050 for Electric Energy and Hydrogen Fuel." *Proceedings* of the 15th Annual Meeting, National Hydrogen Association, Los Angeles, April 27–30, 2004.

[16] P. Kruger, J. Blakeley, and J. Leaver, "Potential in New Zealand for Use of Hydrogen as a Transportation Fuel." *Proceedings* of the 14th World Hydrogen Energy Conference, Montreal, Canada, June 9–13, 2002.

[17] P. Kruger, "Electric Power Required in the World by 2050 with Hydrogen Fuel—Revised." *International Journal of Hydrogen Energy,* 30: 1515–1522, 2005. Text available at www.world-nuclear.org/sym/subindex.htm (for 2004).

[18] U.S. Department of Energy, Energy Information Agency, *International Energy Outlook 2004.* Report DOE/EIA-0408(04). Washington, DC: DOE/EIA, April 2004.

[19] International Atomic Energy Agency, *Energy, Electricity, and Nuclear Power Estimates for the Period up to 2030.* Vienna: IAEA, July 2003.

[20] International Energy Agency, *International Energy Outlook—2002.* Paris: IEA, 2002.

[21] P. Brown, *Hydrogen seen as car fuel of the future. The Guardian,* September 10, 2004.

[22] C. Boardman, A. Hunsbedt, and P. Kruger, *Hydrogen Fuel Demonstration Project for the Los Angeles Air Basin.* General Electric Company and Stanford University Information Meeting. Washington, DC: U.S. Department of Energy, October 21, 1994.

[23] P. Kruger, *Electric Energy: The Potential Showstopper for a Hydrogen Fuel-Cell Fleet.* Seminar Series, California Air Resources Board, El Monte, CA, September 12, and Sacramento, CA, November 4, 2002. Available at www.arb.ca.gov/seminars/sem02/seminars02.htm.

INDEX